# THE ROAD TO RELATIVITY

# THE ROAD TO RELATIVITY

## THE HISTORY AND MEANING OF EINSTEIN'S "THE FOUNDATION OF GENERAL RELATIVITY"

*Featuring the Original Manuscript of Einstein's Masterpiece*

HANOCH GUTFREUND
& JÜRGEN RENN

PRINCETON UNIVERSITY PRESS
PRINCETON AND OXFORD

Library of Congress Cataloging-in-Publication Data
Gutfreund, Hanoch, author.
The road to relativity : the history and meaning of Einstein's "The foundation of general
relativity" featuring the original manuscript of Einstein's masterpiece / Hanoch Gutfreund &
Jürgen Renn.
pages cm
Includes bibliographical references and index.
ISBN 978-0-691-16253-9 (hardcover : alk. paper) 1. General relativity (Physics)—History—
20th century. 2. Einstein, Albert, 1879–1955. Grundlage der allgemeinen Relativitätstheorie.
English. I. Renn, Jürgen, 1956– author. II. Title.
QC173.6.G88 2015
530.11—dc23
2014027854

British Library Cataloging-in-Publication Data is available
This book has been composed in New Century Schoolbook, Minion Pro
Printed on acid-free paper. ∞
Printed in the United States of America
10 9 8 7 6 5 4 3

IN THE LIGHT OF KNOWLEDGE ATTAINED THE
HAPPY ACHIEVEMENT SEEMS ALMOST A MATTER
OF COURSE, AND ANY INTELLIGENT STUDENT CAN
GRASP IT WITHOUT TOO MUCH TROUBLE. BUT THE
YEARS OF ANXIOUS SEARCHING IN THE DARK, WITH
THEIR INTENSE LONGING, THEIR ALTERNATIONS
OF CONFIDENCE AND EXHAUSTION AND THE FINAL
EMERGENCE INTO THE LIGHT—ONLY THOSE WHO
HAVE EXPERIENCED IT CAN UNDERSTAND THAT.

Albert Einstein: *Notes on the Origin of the
General Theory of Relativity*, 1934

# CONTENTS

*A Brief Note on the Publication of This Work*   xi
MENACHEM BEN-SASSON AND MARTIN STRATMANN

*Foreword*   xiii
JOHN STACHEL

*Preface*   xvii

THE CHARM OF A MANUSCRIPT   1

EINSTEIN'S INTELLECTUAL ODYSSEY TO GENERAL RELATIVITY   7

THE ANNOTATED MANUSCRIPT   37

The titles set in italics are Einstein's original titles featured in the manuscript; the page numbers in the titles relate to the annotated pages facing the original facsimile.

*The Foundation of the General Theory of Relativity*
    *A. Fundamental Considerations on the Postulate of Relativity*
        *§1. Observations on the Special Theory of Relativity*
            p. 1: Why did Einstein go beyond special relativity?   39
            p. 2: What was wrong with the classical notions of space and time?   41
        *§2. The Need for an Extension of the Postulate of Relativity*
            p. 3: Why did Einstein see difficulties that others ignored?   43
            p. 4: What was Einstein's happiest thought and how did it come about?   45
        *§3. The Spacetime Continuum. Requirement of General Covariance for the Equations Expressing General Laws of Nature*
            p. 5: Why does Einstein's theory of gravitation require non-Euclidean geometry?   47
            p. 6: What is the role of coordinates in the new theory of gravitation?   49
            p. 7: What is the meaning of general covariance?   51
        *§4. The Relation of the Four Coordinates to Measurement in Space and Time*
            p. 8: What is the geometry of spacetime?   53
            p. 9: When did Einstein realize that gravitation has to be described by a complex mathematical expression?   55
    *B. Mathematical Aids to the Formulation of Generally Covariant Equations*
        *§5. Contravariant and Covariant Four-vectors*
            p. 10: Why tensors, vectors, scalars?   57

p. 11: When did Einstein realize that he needed more sophisticated mathematical methods?   59

§6. *Tensors of the Second and Higher Ranks*
  p. 12: What lessons could be learned from the theory of electromagnetism?   61

§7. *Multiplication of Tensors*
  p. 13: How can tensors be manipulated to produce new tensors by different tensor operations?   63
  p. 13: What were Einstein's heuristic guidelines in his search for a relativistic theory of gravitation?   63
  p. 14: What was Einstein's strategy in constructing a gravitational field equation?   65

§8. *Some Aspects of the Fundamental Tensor $g_{\mu\nu}$*
  p. 15: Why is the metric tensor so fundamental?   67
  p. 16: Why is the Zurich Notebook a unique document in the history of physics?   69
  p. 16: How are volumes measured in curved spacetime?   69
  p. 17: How can a convenient choice of coordinates simplify the theory?   71
  p. 17: What is the difference between a *coordinate condition* and a *coordinate restriction*?   71

§9. *The Equation of the Geodetic Line. The Motion of a Particle*
  p. 18: What is the meaning of a "straight line" in curved space, and how does a particle move under the influence of gravitation?   73
  p. 19: What is the geometric and physical meaning of "Christoffel symbols"?   75
  p. 19: What was Einstein's "fatal prejudice" in the early identification of the gravitational field components?   75

§10. *The Formation of Tensors by Differentiation*
  p. 20: The geodetic line as the "straightest" possible line and its relation to the concept of "affine connection"   77
  p. 21: How do tensors change between neighboring points, or how can one produce new tensors from given tensors by differentiation?   79
  p. 22: What is the geometric context of Einstein's mathematical formulation of general relativity?   81

§11. *Some Cases of Special Importance*
  p. 23: The *Entwurf* theory as an intermediate step toward the general theory of relativity   83
  p. 24: What is the *divergence* of a vector field? What are other vector field concepts?   85
  p. 25: What is the mathematical formulation of energy-momentum conservation in general relativity?   87

§12. *The Riemann-Christoffel Tensor*
  p. 26: What is the geometric meaning of the Riemann-Christoffel tensor?   89
  p. 27: What was the "presumed gravitational tensor" and why was it abandoned?   91

C. *Theory of the Gravitational Field*

§13. *Equations of Motion of a Material Point in the Gravitational Field.*
  *Expression for the Field-components of Gravitation*
  p. 28: When did Einstein lose faith in the *Entwurf* theory?    93
  p. 28: How does a particle move in a gravitational field?    93

§14. *The Field Equations of Gravitation in the Absence of Matter*
  p. 29: What was Einstein's greatest challenge?    95

§15. *The Hamiltonian Function for the Gravitational Field. Laws of Momentum*
  *and Energy*
  p. 30: What is the Lagrangian formalism, and what was its role in the
    genesis of general relativity?    97
  p. 31: What happens to the energy-momentum conservation principle in
    the absence of matter, or can the gravitational field be a source of
    itself?    99
  p. 31: Who was Einstein's main competitor?    99
  p. 32: How can the field equation without matter be generalized to include
    matter?    101

§16. *The General Form of the Field Equations of Gravitation*
  p. 33: The gravitational field equation—at last!    103

§17. *The Laws of Conservation in the General Case*
  p. 34: How is the conservation principle satisfied in a way that Einstein did
    not expect in the early stages of development of his theory?    105

§18. *The Laws of Momentum and Energy for Matter, as a Consequence of the*
  *Field Equations*
  p. 35: Do physical conservation laws follow from symmetries in nature?    107

D. *Material Phenomena*

§19. *Euler's Equations for a Frictionless Adiabatic Fluid*
  p. 36: How do established theories in physics, like hydrodynamics and
    electromagnetism, fit into the new theory of gravitation?    109

§20. *Maxwell's Electromagnetic Field Equations for Free Space*
  p. 37: How did Maxwell represent the laws of electromagnetism by
    mathematical equations and how are these equations affected by
    gravitation?    111
  p. 38: What was the role of "ether" in prerelativity physics, and
    why did Einstein eventually think that space without ether is
    unthinkable?    113
  p. 39: What was von Laue's crucial role?    115

E.

§21. *Newton's Theory as a First Approximation*
  p. 40: How can the validity of the theory be tested experimentally?    117
  p. 40a: What did Einstein wish to clarify and emphasize as an
    afterthought?    119
  p. 41: What does the metric tensor look like in the Newtonian limit?    121
  p. 41: Why was Einstein pleasantly surprised?    121
  p. 42: How could astronomers help confirm certain predictions of the
    theory?    123

*§22. Behavior of Rods and Clocks in the Static Gravitational Field. Bending of Light-rays. Motion of the Perihelion of a Planetary Orbit*
    p. 43: What is the length of rods and the pace of clocks in a gravitational field?   125
    p. 43: Is there a "viable" alternative theory to general relativity?   125
    p. 44: What observation catapulted Einstein to world celebrity status?   127
    p. 45: Explanation of the motion of perihelion of planet Mercury: From disappointment to triumph   129

*Appendix: Presentation of the Theory on the Basis of a Variational Principle*
*§1. The Field Equations of Gravitation and Matter*
    p. A1: Why did Einstein decide not to include this "Appendix" in the printed version of the manuscript "Foundation of General Relativity"?   131
    p. A1: Einstein applies a Hamiltonian (Lagrangian) formulation—different from Hilbert's and different from his own previous one   131
    p. A2: Why did Einstein decide to publish a modified version of this appendix after all? What were the roles of Lorentz and Hilbert?   133
*§2. Formal Consequences of the Requirement of General Covariance*
    p. A3: Is the conservation principle satisfied without any restrictions?   135
*§3. Properties of Hamilton's Function G*
    p. A3: 1916: A year of hard work and new beginnings   135
    p. A4: Einstein acts as a missionary of science   137
    p. A5: Scientific creativity in the midst of personal hardships and national disaster   139

NOTES ON THE ANNOTATION PAGES   141

POSTSCRIPT: THE DRAMA CONTINUES . . .   149

*A Chronology of the Genesis of General Relativity and Its Formative Years*   159

*Physicists, Mathematicians, and Philosophers Relevant to Einstein's Thinking*   165

*Further Reading*   179

*English Translation of "The Foundation of the General Theory of Relativity"*   183

*English Translation of "Hamilton's Principle and the General Theory of Relativity"*   227

*Index*   233

# A BRIEF NOTE ON THE PUBLICATION
# OF THIS WORK

WE CONGRATULATE THE AUTHORS OF THIS BOOK ON THEIR INITIATIVE IN PRESENTING to the nonprofessional reader the history and meaning of Albert Einstein's greatest intellectual achievement—his general theory of relativity. The book is the result of the scholarly effort of its authors, yet their institutional affiliation carries in this context an additional symbolic value. Albert Einstein, The Hebrew University of Jerusalem, and the Max Planck Society form a triangle of relations that deserves some attention.

Albert Einstein was a founder of the Hebrew University of Jerusalem. He served on its Board of Governors and as the first chairman of its Academic Committee. On the occasion of the opening of the university in 1925, he published a mission statement in which he wrote: "A university is a place where the universality of the human spirit manifests itself," and expressed the wish that "our University will develop speedily into a great spiritual center, which will evoke the respect of cultured mankind the world over." This vision has been amply fulfilled.

In 1950, Einstein gave profound expression to his lifelong commitment to the Hebrew University: he bequeathed his own true wealth—his personal papers and literary estate—to the university, making it the eternal home of his intellectual legacy. Today they make up the Albert Einstein Archives, which constitute a cultural asset of supreme importance to mankind. Its holdings are unique—they consist of numerous manuscripts, prolific correspondence, and a large variety of additional material about Einstein. The material in the archives sheds light on the multifaceted aspects of Einstein's scientific work, his political activities, and his private life. The documents have enabled scholars to trace the development of the ideas that led Einstein to his general theory of relativity. It is this intellectual journey that is the subject matter of the present book.

Einstein submitted his theory of general relativity to the Royal Prussian Academy of Sciences in November 1915. In 1917, he became the first director of the Kaiser Wilhelm Institute for Physics. After one of the predictions of the new theory was confirmed, the *Berliner Illustrirte Zeitung*, featuring Einstein's photo on the front page, proudly announced him as "a new celebrity in world history." All this acclaim ended tragically when the Nazis came to power, and Einstein, like many of his colleagues of Jewish heritage, became homeless in his own homeland. After the defeat of Nazi Germany, when the magnitude of suffering inflicted on nations, ethnic groups, and individuals by the Nazi policy and ideology became clearly evident, Einstein rejected numerous invitations and suggestions to return to Germany and to rejoin German scientific institutions. For instance, Einstein was invited to join the newly established Max Planck Society—successor to the Kaiser

Wilhelm Society—by its president Otto Hahn. His refusal was sharp and clear. In 1947, Einstein also refused to approve any publication of his writings in Germany. His stance changed in 1954, when he agreed to the publication of a new German edition of his popular book *The Special and General Theory of Relativity*.

In 1959, the Max Planck Society pioneered academic contacts with the Weizmann Institute in Israel, even before diplomatic relations between Germany and Israel had been established. These contacts were the beginning of a long and productive academic cooperation between the two countries. Currently, Max Planck researchers and their Israeli colleagues are working together on 88 joint projects. Nearly a quarter of these projects involve scientists from the Hebrew University, demonstrating how well our two scientific institutions complement each other. In the recently founded Max Planck–Hebrew University Center on Sensory Processing of the Brain in Action we have joined forces to shed light on the functional building blocks of the brain, the neural circuits.

Regarding the theme of this book, Einstein's theory of general relativity, its consequences, and its history are being explored at several Max Planck Institutes, including the Albert Einstein Institute in Golm and the Max Planck Institute for the History of Science in Berlin. Between 1999 and 2005, the Max Planck Society realized a large-scale historical research project investigating the involvement of its predecessor society in Nazi crimes. To mark the centennial of Einstein's "miraculous year," the authors of the present book collaborated in producing the 2005 Berlin exhibition *Albert Einstein—Chief Engineer of the Universe* on behalf of both the Max Planck Society and the Hebrew University of Jerusalem. The 100th anniversary of the discovery of general relativity has brought our institutions together once more and motivated the authors to produce this book. We are grateful for this enterprise.

PROFESSOR MENACHEM BEN-SASSON
*President of the Hebrew University of Jerusalem*

PROFESSOR MARTIN STRATMANN
*President of the Max Planck Society*

# FOREWORD

You consider the transition to special relativity as the most essential
thought of relativity, not the transition to general relativity. I consider the reverse to be correct. I see the most essential thing in the
overcoming of the inertial system, a thing that acts upon all processes but undergoes no reaction. This concept is, in principle, no better
than that of the center of the universe in Aristotelian physics.[1]

WE ARE AT THE THRESHOLD OF THE CENTENNIAL OF EINSTEIN'S DISCOVERY IN 1915 OF
general relativity, a most fitting occasion for the publication of this volume, an annotated
edition of the paper that is the culmination of the transition from the special to the general theory. Its two authors are eminently qualified for their task: Hanoch Gutfreund par-
ticipated in the publication of an elegant facsimile edition of Einstein's 1912 manuscript
on the special theory of relativity[2] that contains some of the missing links for reconstructing the development of that theory, while Jürgen Renn was a key editor of Einstein's 1912
Zurich Notebook,[3] a crucial document for reconstructing the development of the general
theory.

It must be emphasized that the steps involved in the development of the general theory
of relativity were revolutionary. In many ways, they involved a much greater break with
traditional physics than did the steps leading from Galilei-Newtonian physics to the special theory of relativity. I mention three of them here.

1. First, there is no longer such a thing as an "empty" region of space-time. At the very
least, there is always a chrono-geometric field (metric) regulating the behavior of ideal
rods and clocks, and a compatible inertio-gravitational field (connection) regulating the
force-free motion of material bodies. In the age-old controversy between absolute and
relational concepts of space and time, it now seems difficult to maintain an absolutist
position. As Einstein put it:

> [The metric tensor components] describe not only the field, but at the same time
> also the topological and metrical structural properties of the manifold. . . . There is
> no such thing as empty space, i.e., a space without a field.[4]

Using the following old metaphor,

> Instead of thinking of space and time as a stage, on which the drama of matter
> unfolds, we have to imagine some ultra-modern theater, in which the stage itself
> becomes one of the actors.[5]

step 1 may be put this way: There is no such thing as an empty stage without actors on it.

2. Continuing this metaphor, we can say that these space-time structures no longer form a fixed stage, on which different dramas of matter and fields may be enacted: stage and actors interact. A new drama requires a new stage. Not only is the *local* structure (in the sense of a finite patch) of space-time dynamized, but the *global* structure (in the sense of the entire manifold topology) is no longer given a priori. For each solution to the gravitational field equations given locally, one must work out the global topology of the maximally extended manifold(s) compatible with this local space-time structure.[6]

3. Finally, the stage has no properties of its own that are independent of the action. The same drama cannot be enacted on different parts of the stage: as the actors move about, they carry the stage along with them. Expressed less metaphorically, the points of the bare manifold have no inherent properties that distinguish one point from another; rather, all such distinctions depend on the presence of fields and matter. Many textbooks on general relativity still refer to these bare points as "events," incorrectly suggesting that, as in all previous physical theories, the points are physically individuated a priori, thus obscuring this truly revolutionary feature.

Together, these three steps gave rise to the concept of a background-independent theory: no actors, no anything. Einstein put it this way:

> Space-time does not claim existence on its own, but only as a structural quality of
> the field.

If we consider otherwise empty regions, in which only the chrono-geometrical metric and the corresponding inertio-gravitational connection are present, then the points of such a region cannot be individuated by anything but the properties of these fields.

Recently, I put it this way:

> One of the most crucial developments in theoretical physics was the move from
> theories dependent on fixed, non-dynamical background space-time structures to
> background-independent theories, in which the space-time structures themselves
> are dynamical entities. . . . Even today, many physicists and philosophers do not
> fully understand the significance of this development, let alone accept it in practice.
> One must assume that, in an empty region of space-time, the points have no inher-
> ent individuating properties—nor indeed are there any spatio-temporal relations
> between them—that do not depend on the presence of some metric tensor field. . . .
> Thus, general relativity became the first fully dynamical, background-independent
> space-time theory.[7]

But one must not think that this is the end of the story. In Winston Churchill's immortal words: "Now this is not the end. It is not even the beginning of the end. But it is, perhaps, the end of the beginning."[8] When viewed retrospectively, the 1915 paper is the culmination of what I have called Einstein's Odyssey;[9] but when viewed prospectively, it is the first step in an intellectual journey that is still ongoing. Like many other foundational papers, it functions as a nodal point, summing up the past and opening wide vistas for the future.

Even at the moment of his greatest triumph, Einstein never doubted this. In 1916, he wrote:

> It appears that the quantum theory would have to modify not only Maxwellian electrodynamics, but also the new theory of gravitation.[10]

There is a well-known tension between the methods of quantum field theory and the structure of general relativity. The methods of quantization of non-general-relativistic theories are based on the existence of a fixed kinematical background space-time structure, providing the *where* and *when* for all events. This space-time structure is needed both for the development of the formalism of the dynamical theory to be quantized and—equally important—for its physical interpretation: If a system prepared *here* and *now* is subject to some dynamical interactions, what will be the result of a measurement made on the system *there* and *then*?

General relativity does not fit this pattern. It is a background-independent theory with no fixed, nondynamical structures, and hence it has no kinematics independent of its dynamics. In such a theory, *here* and *now*, and *there* and *then*, are not part of the questions posed to a system but part of the answers given!

However there is hope: general relativity and special-relativistic quantum field theories do share one fundamental feature that often is not sufficiently stressed: the primacy of processes over states. The four-dimensional approach, emphasizing processes in regions of space-time, is basic to both. The ideal approach to quantum gravity would be a background-independent method of quantization that takes process as primary.[11]

The challenge of finding such an approach still awaits solution. But even if or when a satisfactory quantization of Einstein's gravitational field equations is found, that still will not be the end of the story, as Einstein always realized. Early in 1917, he wrote:

> But I do not doubt that sooner or later the day will come, when this way of conceiving [of gravitation] will have to give way to another that differs from it fundamentally, for reasons that today we cannot even imagine. I believe that this process of deepening of theory has no limit.[12]

JOHN STACHEL

## NOTES

1. Einstein to Georg Jaffe, 19 January 1954, cited from John Stachel, *Einstein from 'B' to 'Z'*, p. 294.
2. *1912 Manuscript on the Special Theory of Relativity: A Facsimile* (New York: George Braziller, 1996).
3. Jürgen Renn, ed., *The Genesis of General Relativity*, 4-vol.set, in *Boston Studies in the Philosophy of Science*, vol. 250 (Dordrecht: Springer 2007); *Michel Janssen, John Norton, Jürgen Renn, Tilman Sauer, and John Stachel*, vol. 1: *Einstein's Zurich Notebook: Introduction and Source*; vol. 2: *Einstein's Zurich Notebook: Commentary and Essays.*

4. Albert Einstein, "Relativity and the Problem of Space," in *Relativity: The Special and General Theory, Appendix V* (New York: Crown 1952), p. 155.

5. See Einstein's "Odyssey: His Journey from Special to General Relativity," *The Sciences* 19 (1979): 14–15, 32–34; reprinted in *Einstein from 'B' to 'Z',* pp. 225–232.

6. Criteria must be given for the selection of such an extension, or extensions, if these criteria do not lead to a unique selection.

7. John Stachel, "The Hole Argument," *Living Reviews in Relativity*," http://www.livingreviews.org/lrr-2014 –1, sec. 1, "Why Should We Care?

8. Speech at the Lord Mayor's Day Luncheon at the Mansion House, London, 9 November 1942.

9. See Einstein's "Odyssey: His Journey from Special to General Relativity."

10. "Approximative Integration of the Field Equations of Gravitation," in CPAE, vol. 6, *The Berlin Years: Writings 1914–1917* (English translation supplement), tr. Alfred Engel (Princeton, NJ: Princeton University Press, 1997), 201–209; citation from p. 209.

11. For further discussion of this question, see "The Hole Argument," sec. 6.4, "The Problem of Quantum Gravity."

12. Einstein to Felix Klein, 4 April 1917, cited from CPAE, vol. 8A (Princeton, NJ: Princeton University Press, 1998), 431.

# PREFACE

THIS BOOK PRESENTS A FACSIMILE OF THE MANUSCRIPT OF ALBERT EINSTEIN'S canonical 1916 paper on the general theory of relativity, which may be considered one of the most sophisticated intellectual achievements produced by a single human mind.

Each page of Einstein's manuscript is accompanied by brief essays to guide the nonspecialist through Einstein's arguments and to place this work in a broad intellectual and historical context. The explanatory texts refer to the topics on the specific page and to relevant historical backgrounds. The different kinds of commentaries are differentiated by their typographic styles. So as not to interfere with a fluent reading of the essays, the bibliographic information and suggestions for further reading pertaining to the content of each page are given at the end of the book.

The reproduction of the manuscript is preceded by a comprehensive historical introduction narrating the evolution of general relativity into a full-fledged theory. The introduction and the texts accompanying the manuscript tell essentially the same story but they do so in a different style, in a different format, and sometimes at a different level of exposition. It is hoped that this dual approach will help readers appreciate the development from different angles and will help them choose which track they would like to pursue.

The advantage of presenting this story on the background of Einstein's manuscript is explained in the prologue, "The Charm of a Manuscript." This prologue also explains how the manuscript moved from Berlin, where it was written, to its eternal home at the Hebrew University in Jerusalem, where Einstein's papers are preserved.

The manuscript is followed by a postscript describing the aftermath of the completion of the theory and its immediate cosmological implications. Einstein's 1916 publication does not in fact represent his final views on a number of issues relating to general relativity. A timeline is provided to help orient the reader in the developments between 1905 and 1932, covering the genesis and formative years of general relativity. For the benefit of the reader with a more advanced background in science, the English translation of Einstein's 1916 paper is appended.

Another helpful element is a glossary of scientists and philosophers relevant to Einstein's thinking, featuring their images and brief biographical sketches. This glossary explicitly demonstrates what is conveyed throughout the text: that Einstein maintained a broad network of connections and exchanges with friends and colleagues as he struggled with the challenge of creating his new theory of gravitation. We are grateful to Giuseppe Castagnetti for composing the biographical notes and to Beatrice Hilke for her assistance with the images.

The story outlined in this book is known to few but is of interest to many, concerning as it does one of the most important turning points in the history of science. This book is an attempt to make this development accessible to a broad audience.

We specifically chose to illustrate the text with drawings by Laurent Taudin in a cartoonlike style to add a light, anecdotal flavor. We are grateful to Laurent for his inventiveness but also for his striking capacity to grasp the essence of the subject matter.

We are grateful to Ingrid Gnerlich from Princeton University Press for guiding us through different stages of this project and to the anonymous referees appointed by Princeton University Press for their suggestions, which we have followed. Special thanks are due to our colleagues and friends Jean Eisenstaedt, Robert Schulmann, and Bernard Schutz for critically reading earlier versions of the manuscript. Particular thanks go to our friends Michel Janssen and John Stachel, whose suggestions were very helpful in improving the text.

We are grateful to the staff of the Albert Einstein Archives at the Hebrew University for their assistance, specifically to director Roni Grosz, to Barbara Wolff, and to Chaya Becker. Our thanks also go to Diana Kormos-Buchwald, the general editor of the Einstein Papers Project, for allowing us to quote extensively from the published volumes of the Collected Papers of Albert Einstein, as well as for her personal encouragement with which she has accompanied our work.

This project owes a special debt to two institutions that were directly and indirectly involved. The Hebrew University allowed us to use the manuscript and other archival material, and the Max Planck Institute for the History of Science became the venue where this project was created. We are therefore grateful for the support of both institutions.

Finally, we acknowledge with appreciation and gratitude the invaluable editorial assistance and professional support of Lindy Divarci.

THE ROAD TO RELATIVITY

# THE CHARM OF
# A MANUSCRIPT

MANUSCRIPTS OF IMPORTANT DOCUMENTS IN THE HISTORY OF MANKIND AND OF LET-
ters and writings of known individuals are all available in easily accessible printed form.
Still, the handwritten originals maintain their charm and generate interest and aesthetic
appeal. They are displayed at exhibitions and purchased by collectors at public auctions.
Such originals give us a sense of kinship with the author and a glimpse into his or her
working process. The difference between an original manuscript and its printed version is analogous to the difference between an original work of art and its reproduction,
as described by Walter Benjamin in his book *The Work of Art in the Age of Mechanical
Reproduction*. There he writes: "Even the most perfect reproduction of a work of art is
lacking in one element: its presence in time and space, its unique existence at the place
where it happens to be."[1]

In the early stages of his career, Einstein was not aware of this aspect of his written work
and usually disposed of the original manuscripts as soon as the articles were published
in print. Thus, none of the original manuscripts of his papers from 1905, his "miraculous
year," survive. However, there exists a handwritten version of Einstein's 1905 paper on
the Special Theory of Relativity, "On the Electrodynamics of Moving Bodies."[2] In 1944,
he reproduced it in his handwriting as a contribution to the war effort. It was put up for
auction and raised $6.5 million. It is now held at the Library of Congress.

The earliest extant scientific manuscript is a 70-page-long review article on Einstein's
Special Theory of Relativity. It was written at the request of the editor of the *Handbuch der
Radiologie*, which published annual volumes of review articles on progress in different
fields of science. Owing to delays in publication and the breakout of World War I, this
article was never published. The manuscript remained in the custody of the publisher,
and years later, in 1995, it was offered at auction by Sotheby's, New York. Banker Edmond
Safra bought the manuscript and donated it to the Israel Museum in Jerusalem as a gesture to its illustrious mayor, Mr. Theodor "Teddy" Kollek. The museum framed and hung
each page, exhibiting the manuscript as a work of art, which attracted large audienc-
es.[3]Although most of the visitors did not understand the content, the language, and/or
the handwriting of the manuscript pages, they nevertheless admired and were fascinated
by the exhibition. This is the effect such manuscripts produce.

The manuscript reproduced in this book marks the conclusion of Einstein's intellectual odyssey toward his General Theory of Relativity.[4] About two months after his
final presentation of the theory to the Royal Prussian Academy of Science, he wrote to
Lorentz: "My series of gravitation papers are a chain of wrong tracks, which nevertheless

did gradually lead closer to the objectives. That is why now finally the basic formulas are good, but the derivations abominable; this deficiency must still be eliminated."[5] Without eliminating what appeared to him as an avoidable complexity, Einstein submitted the manuscript for publication to Wilhelm Wien, the editor of *Annalen der Physik*, the leading journal in physics at the time, on March 19, 1916. In the submission letter, Einstein informed the editor that he had also discussed, with the publisher of the journal, an additional publication of this manuscript as a separate booklet. The article "Foundation of General Relativity" was published on May 11th in *Annalen der Physik* and also separately.

The general relativity manuscript is now part of the Albert Einstein Archives at the Hebrew University of Jerusalem. How it got there is a complex story, the details of which are not completely known. Apparently, Einstein gave the manuscript to his friend the physicist astronomer Erwin Freundlich, with whom he had an ongoing dialogue on possible observational tests of phenomena predicted by the new relativistic theory of gravitation. In 1920, Freundlich was one of the founders of the Einstein Donation Fund, which supported the construction of the Einstein Tower in Potsdam, where such tests were to be conducted. We do not know when and why Einstein gave the manuscript to Freundlich. The nature of this "gift" later became a point of dispute between them. By the end of December 1921, the relationship between the two colleagues and friends had deteriorated. Einstein resigned from the board of trustees of the fund and demanded that Freundlich return the manuscript. In an angry letter to Freundlich he wrote:

> As concerns my manuscript, I ask you to arrange to have it handed over to me immediately, without wasting another word on it. I had requested that you send it back to me in the summer. You promised in writing to send it back immediately upon your return from your summer trip. When you did not follow through with it then, my wife wrote you a letter in this regard, to which you did not respond. Now

Erich Mendelsohn:
Sketches for the design of
the Einstein tower, 1918.
bpk / Kunstbibliothek,
Staatliche Museen zu Berlin

you retrospectively contend I had given the manuscript to you, for which there was absolutely no reason. As if this were not enough, you took steps behind my back to sell the manuscript abroad, as you yourself told me. I hope now that you will do your duty without my having to admonish you again. [6]

Einstein retold the story of the manuscript in a letter to Arnold Berliner,[7] editor of the journal *Naturwissenschaften*, who tried to mediate this dispute. Einstein concluded: "I find Freundlich's conduct such that I want nothing to do with him. . . . It no longer concerns the manuscript but the man, whom I cannot trust anymore." The handwritten draft of this letter contains a sentence that Einstein crossed out: "Auf das Manuscript verzichte ich hiermit; mit Freude daran." (I am happy to do without the manuscript.)

Freundlich returned the manuscript, and in April 1922, Einstein entrusted the industrialist and philosopher of science Paul Oppenheim with selling it, giving the following instructions: "The Jewish University of Jerusalem shall be given half of the proceeds; of the remaining half you may dispose as your conscience tells you."[8] Thus, Einstein left it to Oppenheim's discretion to decide on Freundlich's claim to rightful ownership of the manuscript, although in a postscript Einstein stated that he was deeply convinced that Freundlich had no right to it and that his behavior was deceitful. Oppenheim was a friend of both adversaries and did not want to serve as a moral judge between them. Rather, he wished to restore their friendship.

In July 1923, Einstein took another course of action. He asked Heinrich Loewe, a prominent member of "The Preparatory Board of the Hebrew University and the Jewish National Library in Jerusalem" to sell the manuscript. This time the instructions concerning the allocation of the proceeds were very specific: They were to be distributed in equal parts among the library in Jerusalem, the Einstein Donation Fund, the fund securing Mrs. Freundlich's pension, and Einstein himself, who would then donate his share to charity. These instructions were confirmed in a letter from Loewe to Einstein.[9]

The manuscript was not sold, and its fate is revealed in correspondence between Einstein and his wife Elsa when in 1925 he spent two months in South America. Only his letters to Elsa survive; we do not know what she wrote to him. On April 15, in a postscript, he wrote: "Do not give away the manuscript, dear Elsa. . . . The time is not good for selling it. Better after my death."[10] Einstein did not know that on March 19th, Leo Kohn had already received the manuscript from Elsa on behalf of the Board of Trustees of the University of Jerusalem. The document,[11] signed by Kohn, that confirms this transaction stipulates that it be returned "without delay to Professor Einstein, in case any inconvenience be caused to him by the University's acceptance of the manuscript." This document also states that Mrs. Einstein should receive 2000Mk, to be transferred to the Einstein Fund in Potsdam for the use of Prof. Dr. Freundlich, and 400Mk should be given to Mrs. Einstein for her charities.

When Einstein learned that the manuscript was on its way to Jerusalem, he wrote to Elsa, on April 23rd, with relief: "I am glad that I now got rid of the manuscript and thank you for doing me this favor of love (*Liebesdienst*); better than burned or sold."[12]

The general relativity manuscript has been in the possession of the Hebrew University since its opening on April 1, 1925, and is cherished as one of the university's most precious treasures. The manuscript was displayed for the first time in its entirety at an exhibition marking the 50th anniversary of the Israeli Academy of Science. Each one of its 46 pages was enclosed in a box with controlled illumination and microclimate. Like its 1912 predecessor, the manuscript attracted crowds of interested and excited visitors.

In 2013, the European Space Agency launched an Automated Transfer Vehicle (ATV-4), named "Albert Einstein," carrying supplies and equipment to the International Space Station (ISS). The cargo of ATV-4 contained the first page of the manuscript described in this book, which astronaut Luca Parmitano signed on board the ISS as a symbolic gesture

acknowledging the importance of this manuscript and of what it represents in the history of mankind.

This is the story of a single albeit very important manuscript. The Albert Einstein Archives at the Hebrew University contain many such manuscripts, all of which constitute inspiring chapters in the history of physics. They are being edited and explored by historians of science at the Einstein Papers Project at the California Institute of Technology and elsewhere. All shed light on how science was done in the formative years of modern physics.

## NOTES

1. Walter Benjamin, *The Work of Art in the Age of Mechanical Reproduction* (London: Penguin, 2008).
2. Albert Einstein, "On the Electrodynamics of Moving Bodies" (1905), in CPAE vol. 2, Doc. 23, pp. 140–171.
3. The facsimile copy of this manuscript was published by George Braziller as *Einstein's 1912 Manuscript on the Special Theory of Relativity* (New York: Braziller, 1996).
4. It has been analyzed in detail in Michel Janssen, "Of Pots and Holes: Einstein's Bumpy Road to General Relativity," *Annalen der Physik* 14 (2005), Supplement: 58–85; and in Tilman Sauer, "Einstein's Review Paper on General Relativity Theory," in *Landmark Writings in Western Mathematics*, 1640–1940, ed. I. Grattan-Guiness (Amsterdam: Elsevier, 2005), 802–822.
5. Einstein to H. A. Lorentz, 17 January 1916, CPAE vol. 8, Doc. 183.
6. Einstein to Erwin Freundlich, 20 December 1921, CPAE vol. 12, Doc. 330, AEA 11–314.
7. Einstein to Arnold Berliner, 24 December 1921, vol. 12, Doc. 339, AEA 11–318, AEA 11–319.
8. Einstein to Paul Oppenheim, 15 April 1922, CPAE vol. 13, Doc. 146, AEA 11–323.
9. Heinrich Loewe to AE, 30 July 1923, AEA 36–860.
10. Einstein to Elsa Einstein, 15 April 1925, AEA 143–186.
11. Leo Kohn, 19 March 1925, AEA 36–863.
12. Einstein to Elsa Einstein, 23 April 1925, AEA 143–187.

# EINSTEIN'S INTELLECTUAL ODYSSEY TO GENERAL RELATIVITY

Einstein and Newton's apple: Newton's ingenious insight led him to the conclusion that the motion of a falling apple and the motion of the moon orbiting the Earth are both governed by the same physical law—the law of universal gravitation.

EINSTEIN'S FAMOUS 1905 PAPERS SHOOK THE FOUNDATIONS OF CLASSICAL PHYSICS.[1] They challenged the idea of light as a wave, gave striking proof for the existence of atoms, led to a new understanding of space and time, and identified mass as a form of energy. The revolution of space and time that started in 1905 with Einstein's formulation of the special theory of relativity was soon seen to be incomplete. Attempts to fit Newton's well-established law of gravity into the framework of this theory did not succeed, at least not without giving up basic principles of mechanics. Although this problem did not lead to urgent empirical queries, it led Einstein in 1907 to question the special theory's concepts of space and time and caused him to continue the revolution with his 1915 theory of general relativity.

The following remarks introduce the reader to the development of Einstein's ideas and attitudes as he struggled for eight years to achieve a theory of general relativity that would meet the physical and mathematical requirements laid down at the outset. Many of the points discussed here will appear again in the annotations to the specific pages of the manuscript. However, before getting there, we present the whole story as it evolved, with all its dilemmas, wrong paths, misinterpretations, and misunderstandings, to which Einstein himself admitted as he progressed on the bumpy road to his final goal.

## A TALE OF THREE CITIES: PRAGUE, ZURICH, BERLIN

The Einstein specialist John Stachel refers to the development of general relativity as "a drama in three acts."[2] According to this scenario, the first act occurred in 1907 with the formulation of the basic idea to which Einstein referred as the "equivalence principle." The second act took place in 1912, when Einstein realized that the gravitational field is mathematically represented by 10 functions of spacetime coordinates, which form the metric tensor associated with the non-Euclidean geometry of spacetime. The third act, with its "happy end," occurred in November 1915, when Einstein formulated the gravitational field equations and explained the anomalous precession of the perihelion of the planet Mercury.

We propose another script for this dramatic development, which explores its geography. Einstein conceived the notion of the equivalence principle when he was still employed at the patent office in Bern, and he published it in a review article on special relativity in 1907. In that article Einstein also discussed some of its immediate implications, such as the bending of light rays in a gravitational field and the effect of gravitation on the pace of clocks. This can be viewed as a prelude to the real "drama," which began in 1911 when he went to Prague. Then, after a pause of four years, Einstein resumed his interest in gravitation and pursued it intensively—almost exclusively and sometimes obsessively—until his triumphal achievement. It is to this period that we refer as a "Tale of Three Cities." Each of these cities served as a stage for a specific chapter in this development. Each one provided a different social and political environment, and each was characterized by a different phase in his family life. How these circumstances related to his scientific work is discussed in several Einstein biographies.

**PRAGUE** In 1909, Einstein was appointed extraordinary professor at the University of Zurich. For the first time he held a position that carried certain academic and public prestige. Less than six months later, he was offered the even more prestigious position of full professor at the German part of the Charles University of Prague when a vacancy in theoretical physics opened up there. Einstein's candidacy to this position was most strongly supported by Anton Lampa, a professor of experimental physics and ardent follower of Mach, who hoped that Einstein would further promote Mach's ideas.[3]

After some delay and despite the reluctance of his wife Mileva to leave Zurich, where she felt more comfortable, and despite appeals of students to the authorities of the university to make every effort to keep him in Zurich, Einstein accepted the offer and went to Prague in April, 1911.

In Prague, Einstein wrote 11 scientific papers, 6 of which were devoted to relativity. In the first of these papers, published in 1911, he discussed the bending of light and the gravitational redshift, which he had already discovered in 1907,[4] but now Einstein explored them as observable effects. In the Prague papers, he focused on developing a consistent theory of the static gravitational field based on the equivalence principle. Just like Newton's theory of gravity, it involved a gravitational potential represented by a single scalar function, now given by a variable speed of light. Nevertheless, some basic features of the final theory of general relativity had already been conceived by then. Among them was the understanding that the source of the gravitational potential is not only the mass of concrete bodies but also the equivalent mass of the energy of the gravitational field itself. However, until the end of this period Einstein still assumed that the gravitational

"In Prague, I found the necessary concentration for developing the basic idea of the general theory of relativity."

potential was represented by a single function—the space-dependent speed of light—and the theory he developed was restricted to a static gravitational field.

It is interesting to note that Einstein's work on gravitation in Prague was done to a large extent within the context of a controversy with the physicist Max Abraham, famous for his contributions to electrodynamics and electron theory. Abraham was the first to publish, in January 1912, a complete theory of the gravitational field formulated within the framework of Minkowski's four-dimensional spacetime.[5] At first, Einstein was impressed but then reacted skeptically. To his friend Besso he wrote: "At first (for 14 days) I too was completely bluffed by the beauty and simplicity of his formulas."[6] Yet, in the ensuing controversy both Abraham and Einstein developed important insights.

In a foreword to the Czech edition of 1923 of his famous little popular book "About the Special and General Theory of Relativity in Plain Terms," Einstein refers to his work in Prague:[7]

> I am pleased that this small book . . . should now appear in the native language of the country in which I found the necessary concentration for developing the basic idea of the general theory of relativity which I had already conceived in 1908 [he must have meant 1907]. In the quiet rooms of the Institute of Theoretical Physics of Prague's German University in Vinicna Street, I discovered that the principle of equivalence implies the deflection of light rays near the Sun by an observable amount. . . . In Prague I also discovered the shift of spectral lines towards the red. . . . However, the decisive idea of the analogy between the mathematical formulation of the theory and the Gaussian theory of surfaces came to me only in 1912 after my

return to Zurich, without being aware at that time of the work of Riemann, Ricci, and Levi-Civita. This was first brought to my attention by my friend Grossmann.

**ZURICH** In 1911, Marcel Grossmann was appointed dean of the mathematics-physics department of the Swiss Federal Institute of Technology (ETH). One of his first initiatives as dean was to write to Einstein asking if he would be interested in returning to Zurich to join the ETH. Einstein agreed, declining an earlier offer from Utrecht as well as an opportunity to go to Leiden, both of which would have been enticing given the proximity of colleagues such as H. A. Lorentz. Whatever the reasons Einstein had for preferring Zurich over Utrecht or Leiden, at that time it was the right decision. A short time after returning to Zurich in August 1912, he began an intensive and fruitful collaboration with Grossmann that became a landmark in the development of general relativity.

During the Zurich period, Einstein produced three documents that played a significant role in the search for a theory of general relativity: the Zurich Notebook, the Einstein-Grossmann *Entwurf* paper, and the Einstein-Besso manuscript. We shall discuss the contents and significance of these documents in the relevant sections of this account of Einstein's roadmap to general relativity, so we only briefly describe them now.

The Zurich Notebook contains Einstein's notes from the intermediate phase of his search for a relativistic theory of gravitation, when he was exploring, with the help of Grossmann, the concepts and methods of tensor calculus and Riemannian geometry. The notebook consists of 96 pages, not all of them devoted to relativity. Einstein nevertheless gave it the title "Relativität." The notes were written between mid-1912 and the beginning of 1913. Einstein used the notebook from both the front and the back, and his entries meet upside down about a quarter way through. This notebook constitutes a very

"With the help of a mathematical friend [Marcel Grossmann] here [in Zurich], I will overcome all difficulties."

important document in the history of science and is of pivotal importance for our understanding of the origins of the general theory of relativity.[8]

The Zurich Notebook essentially contains the blueprint for the generally covariant theory, but owing to a yet immature physical understanding to be described shortly, Einstein abandoned this theory. Instead, he and Grossmann published the "Outline of a Generalized Theory of Relativity and of a Theory of Gravitation," which has since been termed the *Entwurf* theory from its German title, which means outline.[9] Although this theory did not meet Einstein's initial requirement of general covariance, he convinced himself that this was the best that could be done, and despite this and other shortcomings of the theory, he expressed satisfaction with it until the summer of 1915.

The so-called Einstein-Besso manuscript is a collection of about fifty pages of calculations, about half of them in Einstein's handwriting and the other half in Besso's. These pages contain a calculation of the precession of the perihelion of Mercury based on the field equation of the *Entwurf* theory and a calculation of the metric tensor in a rotating frame of reference.[10]

The Swiss Department of the Interior approved the request of ETH for a full professorship for Einstein. However, it lasted only three semesters. Einstein was in great demand, and the next offer he could not refuse came from Berlin.

**BERLIN** In 1913, Max Planck was elected secretary of the Royal Prussian Academy of Sciences. Shortly after his election, Planck launched a campaign to elect Einstein to the academy. In July 1913, Planck went to Zurich with Walther Nernst to present to Einstein a tempting three-part proposal: election to the academy with generous financial support,

"She [Else Löwenthal] was the main reason for my coming to Berlin, you know."

directorship of the Kaiser Wilhelm Institute of Physics without a real administrative burden, and a professorship at the University of Berlin without teaching obligations.

Einstein accepted the offer, giving different reasons to different people in justifying his decision. To Lorentz he wrote: "I could not resist the temptation to accept a position in which I am relieved of all responsibilities so that I can give myself over completely to rumination."[11] But to his good friend Heinrich Zangger he admitted that the main reason for accepting this offer was that this would bring him close to his cousin Elsa, whom he was passionately courting at that time and who would later become his second wife: "Despite being in Berlin, I am living in tolerable solitude. But here I have something that makes for a warmer life, namely, a woman whom I feel closely attached to. . . . She was the main reason for my coming to Berlin, you know."[12]

In November 1913, His Imperial and Royal Majesty Wilhelm II confirmed Einstein's election as a regular member of the physics-mathematics section of the academy. Thus, at the age of 34, he became the youngest-ever member of the academy.

Shortly after Einstein's arrival in Berlin, World War I broke out. Confronted with the realities of war, he eventually left the ivory tower of science to become a political opponent of Germany's involvement in the war. In Berlin, Einstein encountered the phenomenon of anti-Semitism and became aware, more than ever before, of his Jewish identity.[13] In Berlin, his relations with Mileva deteriorated to the point of separation—Mileva and the children returned to Zurich. In the midst of all this, Einstein ardently pursued his scientific work and, according to his own testimony, worked harder than ever.

Einstein continued to work on his and Grossmann's *Entwurf* theory of gravitation and suggested new arguments to support its validity. His satisfaction with the *Entwurf* theory solidified to the point that he was ready in October 1914 to summarize it in a review article, "The Formal Foundation of the General Theory of Relativity,"[14] which he published in the meeting reports of the Royal Prussian Academy of Sciences. It took him less than a year to regret it.

Einstein's doubts concerning the *Entwurf* theory began to build in the summer of 1915. He finally abandoned the theory and, in an outburst of creativity and hard work, completed in November of that year his general theory of relativity.

Einstein had joined Max Planck, Walther Nernst, and many others in Berlin, which at the time was the world capital of physics. Even during the hardships of the war years, the city maintained an inspiring atmosphere and work routine in the physics community. Gerald Holton, a pioneer of Einstein scholarship in its historical and philosophical context, addressed the question,[15] "How much did these facts contribute to Einstein's unique ability to develop, between 1915 and late 1917, his general relativity theory in Berlin? Could he have done so if he had accepted a grand offer from a city in another country?" Holton's clear answer is, "No other man than Einstein could have produced General Relativity, and in no other city than in Berlin," albeit not without help from his friends in Zurich!

## THE CHALLENGE OF GRAVITATION

The 1905 theory of relativity had established a new understanding of space and time, and all physical interactions needed henceforth to fit within its framework. In addition, the theory had combined the laws of conservation of energy and momentum into a single

Even when the train is moving, the coffee does not miss the cup. This is the classical principle of relativity.

law and it had demonstrated that mass is a form of energy. The consequences of this theory could be conveniently described in the framework of a new mathematical formalism developed by Herman Minkowski, Einstein's former teacher at the ETH in Zurich. This formalism[16] combines space and time into one entity—spacetime—and assigns a geometric distance between any two physical events that occur at different positions and different times. One usually refers to points in spacetime as events because they are characterized by location and time of occurrence. The square of this distance is simply the square of the time separation between the two events minus the square of their spatial separation. Observers moving at constant velocity with respect to each other may compute this value using their respective positions and time measurements, and they will get the same result. In other words, Minkowski's four-dimensional spacetime is equipped with a "metric" instruction that is employed to measure the distance between events. This may be compared with the familiar metric instruction to measure the distance between two points in three-dimensional space: sum the squares of the Cartesian coordinate separations.

It was not difficult to adapt the domain of electromagnetism to the new spacetime framework of the theory of special relativity, which had actually been inspired by Maxwell's electrodynamics. But gravitation, that is, the force of gravity between two masses, presented problems in this respect. Because Newton's law of gravity assumes an instantaneous action at a distance, this law in its classical form was not directly compatible with the special theory of relativity. One of the consequences of this theory is that no physical effect can propagate with a speed exceeding that of light in a vacuum. Thus, a new gravitational theory was needed, but it was not clear how such a theory should look, what heuristic assumptions could be made, and even what specific criteria it should satisfy.

But, there was an obvious way to make classical gravitational theory formally compatible with the principles of the special theory of relativity, and Einstein initially pursued this line of thinking. However, the problem with this obvious generalization was that the resulting theory of gravitation seemed to violate Galileo's principle that all bodies fall with equal acceleration. This is one of the basic principles of classical physics, mythically established by Galileo dropping material objects from the top of the tower of Pisa.

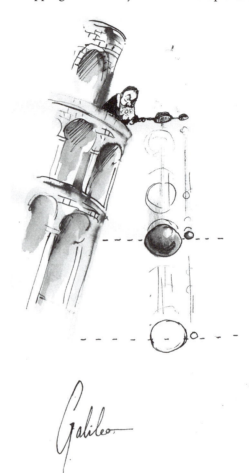

They all fall in the
same time!

Galileo's principle stipulates that the acceleration of free fall is the same for all bodies. Newton accounted for this principle by setting inertial mass equal to gravitational mass. The inertial mass determines the acceleration of a body caused by a given force, while the gravitational mass determines the force exerted on a body by a given gravitational field. The dependence of inertial mass on energy in special relativity must imply that in a relativistic theory of gravitation, the gravitational mass of a physical system should also depend on the energy in a precisely known way so as to maintain Galileo's principle. Einstein concluded that if the theory did not achieve this in a natural way, it was to be abandoned. Contemporary scientists such as Max Abraham and Gustav Mie were quite ready to abandon Galileo's principle in order to obtain a relativistic theory of gravitation in the sense of special relativity.

Einstein's generalization of relativity theory began with Ernst Mach's philosophical critique of classical mechanics.[17] Mach, whose works the young Einstein admired, had claimed that the concept of motion or even of inertial mass can by no means be applied to a single body in absolute space, as claimed by Newton. Instead he suggested that all of classical mechanics should be rewritten in terms of relative motions of bodies and also that the concepts *inertial mass* and *inertial system* should be redefined in this fashion. Centrifugal forces in particular, which cause the curvature of the water level in a rotating bucket, were to be interpreted not as an effect of an accelerated motion with respect to absolute space, as Newton contended in his famous bucket experiment, but as an effect of the presence of other bodies in the universe. Einstein noted that Galileo's principle must somehow be related to Mach's special view of mechanics, which rejects the privileged role of inertial frames of reference as well as the concept of acceleration relative to absolute space. Against this background, Einstein realized that the question of how to maintain Galileo's principle must be answered within the framework of a generalization of the principle of relativity. In short, in the conflict between classical mechanics and special relativity, Einstein in 1907 decided to keep the principle of equivalence of gravitational and inertial mass, and in turn was willing to accept that the theory of gravitation would lie beyond the scope of special relativity.[18]

With the obvious concept of developing a new theory of gravitation within the framework of the special theory of relativity in mind, he concluded: "It turned out that, within the framework of the program sketched, this simple state of affairs could not at all, or at any rate not in any natural fashion, be represented in a satisfactory way. This convinced me that within the structure of the special theory of relativity there is no niche for a satisfactory theory of gravitation."[19]

In hindsight, we know that Einstein was right. There are two distinct types of spacetime structures: one is connected with the classical mechanics of Galileo and Newton, while the other is connected with special relativity, determining the behavior of measuring rods and clocks and thus "chrono-geometry." As John Stachel has emphasized, any theory of gravitation incorporating the equivalence of inertial and gravitational mass must start from an inertio-gravitational field governing the behavior of free particles. Even at the Newtonian level, gravitation can thus be described not as an external force acting on bodies but as a modification of the hitherto-fixed inertial structure of spacetime. The challenge of creating general relativity was thus to establish compatibility between these two structures.[20]

## EINSTEIN'S HEURISTICS: THE EQUIVALENCE PRINCIPLE

If the new theory of gravitation was to include Galileo's principle, it would have to be a generalized theory of relativity, because it would have to give accelerated motion the same status as inertial motion, considering gravitational and inertial forces on a par. What Einstein eventually achieved with general relativity was in fact not so much a further generalization of the relativity of motion but a "relativity" of gravity, which became integrated with inertia to constitute one unified inertio-gravitational field.[21] The relativity theory of 1905 privileged uniform motion as that of a train moving with constant velocity. Thus, the laws of physics have to take the same form in reference frames moving uniformly with respect to each other. Galileo's principle suggested that this might hold true even for reference frames accelerated with respect to each other, because all bodies in such reference frames behave in the same way; namely, they fall in the same time. But to compare an accelerated reference frame with a reference frame at rest and to claim that they are somehow equivalent, one must introduce an additional assumption. In an accelerated reference frame somewhere in empty space, like a spaceship far from Earth, bodies will fall to the ground because of the acceleration. In a reference frame at rest on Earth, bodies will fall to the ground because of Earth's gravitation. If the behavior is the same in both cases, gravitation and the apparent forces in the rocket due to its accelerated motion—also known as inertial forces—must be equivalent. This is Einstein's famous equivalence principle, one of the most important heuristic clues in constructing a generalized theory of relativity. In retrospect, he referred to this idea as "the happiest thought" of his life.[22] The equivalence principle states that the gravitational field has only a relative existence, because for an observer falling freely from the roof of a house there temporarily exists, at least in his or her immediate vicinity, no gravitational field. In particular, all physical processes in a uniform and homogeneous gravitational field are equivalent to those that occur in a uniformly accelerated system of reference without a gravitational field. This concept can be illustrated either by an accelerated spaceship or by the thought experiment of a falling elevator.

Including inertial forces in the attempt to construct a new theory of gravitation had far-reaching consequences. Inertial forces are fictitious forces acting on masses in

Centrifugal Forces—

A "fictitious" force makes it difficult for Einstein to keep his hat on.

accelerated frames of reference, like the centrifugal force experienced on a merry-go-around. Einstein used different types of inertial forces as test cases for the new theory, for example, those in an accelerating spaceship. He also considered inertial forces that act in a rotating system of reference, such as the force shaping the surface of a liquid in a rotating bucket, which Newton used to demonstrate the concept of absolute motion. These "fictitious" forces are actually real forces, but their origin has remained enigmatic in classical physics because they have been ascribed to a mysterious property of absolute space. Considering such inertial forces on a par with the well-known Newtonian force, Einstein could then draw qualitative conclusions as well as derive requirements for the mathematical apparatus of his new theory.

One crucial conceptual insight provided by his thought experiments concerns the bending of light in a gravitational field and the nature of time. Einstein inferred the bending of light rays in a gravitational field from the argument that the path of a light ray in an accelerated laboratory must be curved due to the superposition of the motion of the laboratory and the motion of the light. The conclusion that this result must also be valid in a gravitational field was in agreement with the assumption that energy has not only inertial but also gravitational mass, so that light should be subject to attraction by gravity. The deflection of light in a gravitational field suggests that the speed of light should no longer be assumed to be constant, contrary to special relativity. This qualitative conclusion was supported by an analysis of time synchronization in an accelerated reference frame, as described by Einstein in an article written in 1907 (see note 4). His analysis implied that accelerating clocks at different locations run at different rates. He reached the same conclusion by comparing the rate of clocks located at different positions on a rotating disk.

What causes the curvature of the surface of water in a rotating bucket?

## GEOMETRY ENTERS PHYSICS

The inclusion of rotating reference frames presented another conceptual challenge. Einstein and Max Born had encountered this challenge in 1909 in connection with the special theory of relativity. Paul Ehrenfest had also found independently that, according to special relativity, rods that are used to measure the circumference of a rotating disk should experience a so-called "Lorentz contraction." [23] Therefore, more rods are needed, and the circumference will appear longer than that of a disk at rest. However, rods used to measure the radius of the rotating disk will be unchanged, being perpendicular to the direction of motion. Therefore, the ratio of a rotating disk's circumference to its radius will have to be greater than the value determined in Euclidean geometry when both distances are measured in the frame of reference in which the disk is a rest. This difficulty became known as the "Ehrenfest paradox" and led to controversial discussions. Most participants in this debate considered this problem to be primarily a problem of the definition of a rigid body. However, Einstein identified the Ehrenfest paradox as a key issue to be addressed in seeking a generalization of the theory of relativity. In an article published in 1912, he argued that the ratio of circumference to diameter of a disk in a rotating laboratory is no longer given by $\pi$, indicating that general relativity implies a departure from Euclidean geometry. [24]

What if the world is
intrinsically bent?

   In Einstein's thought process, the equivalence principle and the use of accelerated laboratory models became subordinate to a newly formulated heuristic principle: the principle of general relativity. According to this principle, the new theory of gravitation should admit reference systems in arbitrary states of motion, and it should describe the inertial forces occurring therein as the action of a generalized dynamic gravitational field. This principle and the conceptual changes implied by the accelerated elevator and rotating bucket models played a crucial role in considering the kind of mathematics to be used in formulating the theory of gravitation. Einstein had realized that it would be necessary to go beyond Euclidean geometry. The desire to include arbitrary systems of reference gave him the idea in the summer of 1912 to construct the new theory of gravitation using a generalization of the Gaussian theory of curved surfaces, but he first had to generalize this theory to the four-dimensional world of the theory of relativity. Mathematicians like Bernhard Riemann, Elwin Christoffel, and Tullio Levi-Civita had provided the important background for this generalization, but Einstein was not familiar with their works and had to acquire this new mathematics gradually with help from his friend Marcel Grossmann.

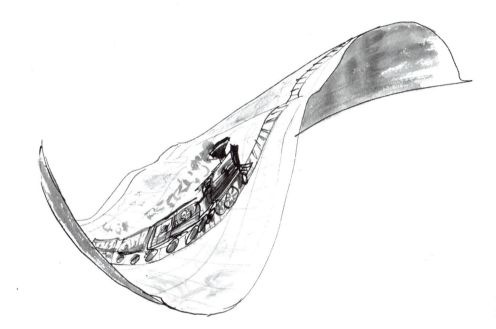

This is how a straight path looks on a curved surface.

The mental model of motions along curved surfaces that was familiar from the world of classical physics also pointed directly to a solution to the problem of determining the equations of motion in an arbitrary gravitational field. An object that is constrained to move along a two-dimensional frictionless curved surface with no other forces than those exerted by the surface itself will always move along the shortest path, called a *geodesic*. This is the simplest generalization of a straight line. The idea could immediately be transferred to the case of motion observed from an arbitrarily accelerated system of reference, corresponding to motion in a gravitational field in the absence of any other forces. Such motion can also be represented as a four-dimensional spacetime geodesic in the curvilinear coordinates used to describe such a system of reference. (Curiously, however, the trajectory described by a freely moving object turns out to be the longest possible path between two given points in spacetime. This is a consequence of the peculiar mathematical properties of the spacetime metric.)

The revised description of the action of gravity meant that the gravitational field was no longer considered to be a force in the sense of Newtonian physics but as the embodiment

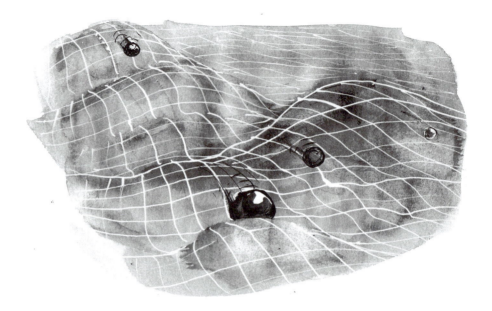

"Spacetime tells matter
how to move; matter tells
spacetime how to curve."
(John Archibald Wheeler)

of geometric properties of a generalized spacetime continuum. The concept of a metric as a generalization of the concept of distance has already been introduced. Whereas a flat surface is characterized by a metric that behaves in the same way everywhere on the surface, the geometric properties of a curved surface must be described by a variable metric. Such a metric associates different actual distances with a given coordinate distance at different locations on the surface. This variable metric turned out to be a suitable representation of the gravitational potential.

### EINSTEIN'S HEURISTICS: A PLAN OF ACTION

In his search for a relativistic theory of gravitation, Einstein could orient himself using a model very familiar to contemporary physicists, because it represented one of the great successes of nineteenth-century physics, namely, the unified theory of all electromagnetic interactions established by James Clerk Maxwell and Heinrich Hertz. It was in fact a remarkable feature of this theory that it did not describe electric and magnetic fields separately but as components of a unified electromagnetic field. This theory was developed into its definitive form by the Dutch physicist Hendrik Antoon Lorentz, who later became one of Einstein's mentors. The central concept of this theory was that of "field." In contrast with describing the interactions of particles due to forces acting at a distance, a field theory is not restricted to the interacting particles but extends to their complete surroundings. Field theory describes how the space-filling field is generated by charges and currents, considered to be the "source" of the field, and it also describes how this field in turn determines the motion of charged particles. A mathematical representation of the physical processes interpreted according to this "Lorentz model" therefore necessarily includes two parts:

- an equation of motion, describing the motion of charged particles in a given electromagnetic field; and
- a field equation, describing the electromagnetic field generated by its sources: charges and currents.

Fortuitously, this theory could be most elegantly presented in the four-dimensional spacetime formulation of Einstein's theory of special relativity, which then became a springboard for the generalization of his theory. A consequence of this reformulation was the insight that the electromagnetic field could manifest itself differently in terms of electric or magnetic fields depending on the state of a physical system. This remarkable property was, in fact, one of the starting points of Einstein's work on the special theory of relativity.

In constructing a relativistic theory of the gravitational field, Einstein was guided in almost every respect by the Lorentz model, including the complementary roles of the gravitational and inertial aspects of the field. These were analogous to the interplay between electric and magnetic aspects of the electromagnetic field. Thus, the new theory would include two parts:

- an equation of motion, describing the motion of particles in a given gravitational field; and
- a field equation, describing the gravitational field generated by its sources: matter and energy.

Einstein solved the first equation in the summer of 1912. The second one posed the greater challenge. The right-hand side of the field equation represents the source of the field or of the potential, and the left-hand side describes, by means of a specific mathematical procedure—a so-called differential operator—how the source generates the field or the potential.

Einstein accepted the Lorentz model as a heuristic guideline, yet he soon found that the task of finding a field equation was the most difficult challenge he had to face in his struggle to formulate a relativistic theory of gravitation. First, he was confronted with the problem of finding an appropriate mathematical object to represent the gravitational potential. Second, the gravitational field equation that would replace the corresponding equation of classical physics had to be compatible with results from both classical gravity and special relativity.

Einstein had to keep in mind that the nature of the gravitational field under normal circumstances, namely, in the case of weak static fields (the Newtonian limit), was well known and satisfactorily described by Newton's law of gravity. Therefore, the relativistic field equation for gravitation had to give the same results as Newton's laws under these circumstances. This constraint may be called Einstein's "correspondence principle." Clearly, the new field equation should also be consistent with the unquestionably valid laws of conservation of energy and momentum in physical interactions. This requirement may be called the "conservation principle." In addition, Einstein's earlier research, guided by the equivalence principle, had yielded a number of discoveries that had to be reproduced in the new theory.

Thus, Einstein's plan of action was to construct a theory that would satisfy the following principles:

- the correspondence principle;
- the conservation principle; and
- the equivalence principle.

In addition, the theory had to be generally covariant.

## THE TWO STRATEGIES: MATHEMATICAL VERSUS PHYSICAL

Even before Einstein could begin to implement his plan of action, he had to solve another problem, namely, how could the gravitational potential be mathematically represented in the context of the relativistic theory he was about to construct? [25] The decisive hint came from his exploration of the equation of motion and from considering the strange properties of rotational motion in such a theory. In 1912 Einstein realized that the gravitational potential is not, as in Newton's theory, given by a simple function but, surprisingly, by a whole array of such functions of space and time that together form a complex mathematical object known as the *metric tensor*. Einstein also realized that this metric tensor is related to non-Euclidean geometry, so his new theory of gravitation would become a theory of the curvature of space and time. This realization constitutes what Stachel refers to as the "Second Act" in the development of the general theory of relativity.[26]

In view of the complexity of the mathematical object representing the gravitational potential, the search for a relativistic field equation turned out to be an extremely challenging research process, in the course of which some of the basic structures of knowledge that dominated Einstein's heuristics had to be revised. His efforts in this direction in the years 1912–1915 can be described as an interplay between two complementary heuristic strategies: a "physical strategy" and a "mathematical strategy."

Regarding the physical strategy, Einstein started with a field equation that, from the outset, gave the correct law of gravitation in the classical Newtonian limit and thus satisfied his correspondence principle. He then modified it so as to render valid the remaining fundamental laws of physics, including the principle of conservation of energy and momentum. The final step was then to find the degree to which this candidate field equation satisfied a generalized relativity principle.

Pursuing a complementary mathematical strategy, Einstein started from a mathematically plausible field equation that would immediately satisfy the most general principle of relativity. He was able to draw on some of the mathematical resources brought to his attention by his mathematician friend Marcel Grossmann, in particular, the "absolute differential calculus" formulated by Ricci and Levi-Civita in a paper published in 1901, in which the previous work of Riemann, Christoffel, and others was developed into a complete calculational scheme. This absolute differential calculus was unknown to physicists for a long time, and in 1912 Einstein most probably knew only very little about it. But after he moved from Prague to Zurich, the contact with Grossmann gave Einstein access to these mathematical methods.

Once a mathematically plausible field equation was established, it remained to check whether it fulfilled the other physical requirements. The serious disadvantage of this procedure was that the relation such an abstract mathematical object bore to the familiar physical knowledge was not initially clear. A systematic alteration of the initial mathematical candidate was therefore necessary with the aim of making its consistent physical

Mathematics or physics—
where to start? That
is the question.

interpretation possible, and this became a necessary part of the strategy. In particular, a candidate field equation had to fulfill the demand that Newton's theory could be recovered for the special case of a weak static gravitational field and assure the conservation of energy and momentum. Furthermore, it had to fulfill the condition that even after it was modified to fulfill these demands, the group of admissible coordinate transformations still remained wide enough to include at least transformations to accelerated reference frames representing the special cases of uniform acceleration and uniform rotation.

The issue of the Newtonian limit of general relativity is complicated by the fact that there are actually two approaches: one going through the intermediate stage of special relativity, and the other via a generalization of gravitational fields within Newtonian physics, allowing slow-motion and quasi-static solutions to be treated. The latter, however, demands a reformulation of Newtonian theory—including the equivalence principle—in terms of mathematical concepts that were introduced only much later by the French mathematician Elie Cartan in reaction to the work of Levi-Civita and Weyl. Before this sophisticated mathematical approach was developed, Einstein was compelled to introduce assumptions about the Newtonian limit that later turned out to be problematic.

## THE BEST THAT CAN BE DONE: THE RISE OF THE *ENTWURF* THEORY

Einstein's collaboration with Grossmann is best reflected in the Zurich Notebook, which eventually led to the publication of the so-called *Entwurf* theory. The problem at the core of Einstein's research documented in this notebook was to find a field equation for the gravitational field, namely, to find a relation determining how this field is generated by its source, that is, by energy and matter. The notebook contains an important entry regarding the assistance of Marcel Grossmann, who referred Einstein to a key mathematical concept, the so-called Riemann tensor, and thereby showed him the royal road to general relativity from the point of view of contemporary theory. Einstein and Grossmann, however, abandoned this royal road soon afterward.

There were two main reasons for this abandonment. They realized that to achieve the correct Newtonian limit they had to impose certain conditions on the choice of admissible coordinates. Grossmann and Einstein also found that requiring their theory to satisfy the conservation of energy and momentum imposed further restrictive coordinate conditions. As a result, they came to the conclusion that a theory based on the Riemann tensor could not be brought into harmony with these physical requirements. Only in 1915, after Einstein had gained a much deeper understanding of how these requirements were to be articulated for the case of his new theory of gravitation, could this conclusion be revised.

Finding the correct gravitational field equation was not just a matter of determining the right mathematical expression but also of integrating mathematical formalism and physical meaning. Just mastering the mathematics would be like using a language in a grammatically correct way without, however, understanding the meaning of its words. The Zurich Notebook reveals Einstein struggling with a new mathematical language into which he was attempting to translate some of the familiar physical knowledge while simultaneously trying to discover the new physical insights it harbored.

Einstein repeatedly alternated between physical requirements derived from Newton's theory of gravity and other conditions suggested by a mathematical formalism appropriate for the description of a curved spacetime, no doubt hoping that the two strategies must ultimately converge. In the notebook, however, he failed to fully realize his aspirations for either one.

Paradoxically, the main result of Einstein's experiments with the mathematical strategy was the more or less successful implementation of the physical strategy. Clearly, the vantage point represented by Newton's classical theory of gravity took precedence over the more speculative new insights connected with the generalization of the relativity principle. It therefore seemed more plausible to Einstein to build on this secure vantage point, even if that meant renouncing some of his loftier ambitions concerning a generalization of the principle of relativity.

After many aborted attempts, Einstein eventually derived, at the end of the Zurich Notebook, a field equation that became known as the core of the *Entwurf* theory. It primarily satisfied the principles rooted in classical physics, namely, the correspondence principle and the conservation principle. Einstein realized that the class of coordinate systems in which the *Entwurf* equation takes on the same form does not satisfy the generalized principle of relativity in the way he imagined. He therefore abandoned with a heavy heart the realization of general covariance. Nevertheless, he could reassure himself that this equation was acceptable because the necessary restriction of the admissible coordinate systems could apparently be justified by the requirement to implement the conservation principle. So there seemed to be a cogent reason for the limited extent to which the generalized principle of relativity was fulfilled in the *Entwurf* theory. From the modern perspective, this theory is incorrect, but at that point Einstein assumed that it was the best that could be achieved. The culmination of this process was the publication in 1913, together with Grossmann, of the article "Outline (*Entwurf*) of a General Theory of Relativity and of a Theory of Gravitation." This article consists of two parts: a physical part authored by Einstein and a mathematical part authored by Grossman.

Initially, Einstein was not completely satisfied with the published theory. In a letter to Lorentz, he referred to the lack of general covariance as an "ugly dark spot" of the *Entwurf* theory.[27] But when Einstein looked for reasons to defend this deficiency, he concluded that the restricted covariance was a necessity. At first he thought he could use energy-momentum conservation to justify the lack of general covariance of his new theory. In December 1913, Einstein wrote to Mach: "The reference system is, so to speak, tailored to the existing world with the energy principle, and loses its nebulous aprioristic existence."[28]

Einstein eventually realized, however, that this argument was not sound. But meanwhile, in the summer of 1913, he had found another, more profound, argument—the famous "hole argument"—claiming that generally covariant theories are bound to violate causality. In the argument's original formulation, Einstein considered a spacetime filled with matter except in a closed region, the hole. Adopting the seemingly plausible assumption that spacetime points can be identified by coordinates, he could show that a specific matter distribution outside the hole does not uniquely determine the gravitational field within. Einstein thought that this result was sufficient to reject all generally covariant theories. Only in late 1915 did Einstein realize that this seemingly reasonable assumption was untenable in his new gravitation theory, because coordinates have no physical significance. The hole argument and its refutation eventually became the starting point for formulating the important concept of background-independent theories, that is, of theories for which time and space are not a fixed stage for the drama of physics.[29]

In 1913, however, it was precisely the erroneous hole argument that motivated Einstein to further consolidate the *Entwurf* theory, whose main "ugly dark spot" seemed to have been overcome. He concluded that "[t]he fact that the gravitational equations are not generally covariant, which still bothered me so much some time ago, has proved to be unavoidable; it can easily be proved that a theory with generally covariant equations cannot exist if it is required that the field be mathematically completely determined by the matter."[30]

From 1913 to late 1915 Einstein was convinced that the now-forgotten *Entwurf* theory constituted the solution to the problem of a relativistic theory of gravitation. To his friend Besso he wrote in 1914: "Now I am completely satisfied and no longer doubt the correctness of the whole system, whether the observation of the solar eclipse succeeds or not. The sense of the matter is too evident."[31]

## THE VARIATIONAL METHOD ENTERS THE SCENE

A nagging doubt, however, persisted: what was the relation of the *Entwurf* field equation to the mathematical tradition of absolute differential calculus? Since his collaboration with Grossmann, Einstein was quite familiar with this tradition, and he knew that the

Riemann and the Ricci tensors would have been the correct mathematical objects on which to build his theory. He was convinced there must be a relation between his *Entwurf* theory and this mathematical language, but it was unclear what that relation might be. To clarify this issue, he again turned to Grossmann for help. In early 1914, the Zurich mathematician Paul Bernays suggested to Einstein and Grossmann that they derive the *Entwurf* field equation from a variational formalism that traces the evolution of a single function known as the Lagrangian. In this formalism, energy-momentum conservation emerges as a natural by-product. Einstein and Grossmann succeeded in finding a Lagrangian function from which the *Entwurf* field equation could be derived, and they observed that this function is invariant under transformations between coordinate systems specified solely by the requirement of energy-momentum conservation. Einstein and Grossmann finally published an article in which they showed how the *Entwurf* equation as well as energy-momentum conservation can be obtained from such a variational formalism.[32] In a sense, this amounted to an adaptation of the mathematical strategy to the *Entwurf* theory. Einstein convinced himself, erroneously, that the variational method he used to derive the field equation led uniquely to the *Entwurf* theory.

## THE EINSTEIN-BESSO MANUSCRIPT: EINSTEIN FAILS TO SEE THE WRITING ON THE WALL

An important touchstone for any new theory of gravitation is not only its capacity to reproduce the laws of planetary motion as established by Kepler and Newton but also to explain the small deviations from these laws as they become evident in particular by the precession of the perihelion of Mercury's orbit. This precession consists in a slight rotation of the ellipse of Mercury's orbit. Most of this rotation can be explained by the influence of other planets and accounted for by Newtonian theory. But at that time astronomers had already known for 50 years that there is a discrepancy between the observed value of this precession and Newtonian theory by 43″ (arc seconds) per century.

As early as 1907, in a letter to Conrad Habicht,[33] Einstein identified the explanation of this discrepancy as one of the goals of a new theory of gravitation. Six years later, he could use this discrepancy to test the *Entwurf* theory. In 1913, together with his friend Michele Besso, Einstein found an ingenious method for approximately solving the *Entwurf* field equation and thus for determining the shifting of Mercury's orbit. This method is described in the Einstein-Besso manuscript, comprising some fifty pages strewn with calculations.[34] This procedure generated the disappointing result of 18″ per century. Nevertheless, it did not, at least not at that time, cause Einstein any doubts regarding the validity of the *Entwurf* theory. Had Einstein taken this result seriously, he could have dismissed the theory and embarked on the right journey two years earlier.

The Einstein-Besso manuscript contains another important calculation that could have led to a similar consequence. It derives the metric tensor in a rotating frame of reference and shows that this is a solution of the gravitational field equation generated by a rotating disk of mass (distant stars).[35] Einstein was very happy with this result because it seemed to confirm Mach's interpretation of Newton's rotating bucket experiment and validated the notion of "rotation at rest" in the relativistic theory of gravitation. Actually, however, Einstein made a mistake in this calculation, which he realized only about two years later, in September 1915.

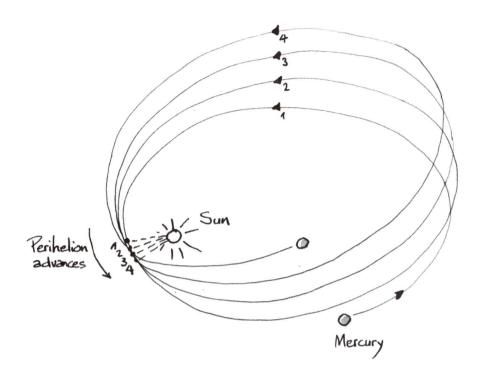

What causes the precession of the perihelion of Mercury's orbit?

## THE *ENTWURF* THEORY: ABANDONMENT

Another central problem on the road to the correct theory, which was solved in the course of elaborating the *Entwurf* theory, was that of satisfying the correspondence principle. This problem arose, as was pointed out earlier, in the inquiry into the relation between the limiting case of the mathematically motivated candidate and classical Newtonian theory. The solution emerged from Einstein's attempt to assess the astronomical consequences of the *Entwurf* theory and to check whether it could explain the perihelion motion of Mercury.

By the fall of 1915, Einstein had decided to abandon the *Entwurf* theory and to revisit his previous inquiries into a covariant theory of gravitation. In retrospect, Einstein gave three reasons for this move:

- The theory could not explain Mercury's perihelion rotation.
- The theory did not confirm Einstein's Machian heuristics (treatment of a rotating system of reference as equivalent to a system at rest).
- The conclusion that the variational method led uniquely to the *Entwurf* field equation turned out to be wrong.

Remarkably, the *Entwurf* theory initially survived all these problems. Even the last one did not lead to the theory's refutation but only to a successful attempt to repair this derivation on a technical level by resorting to a physical argument. The flaw in the hole argument is not listed here because Einstein recognized the fallacy of this argument only after he had completed his general theory of relativity.

The failure of the derivation of the *Entwurf* theory from mathematical principles nevertheless had far-reaching consequences with regard to Einstein's reflection on the results he had achieved. This failure showed that the adaptation of the mathematical strategy to the *Entwurf* theory by means of a variational formalism definitely did not single out this theory as the only possible one, as he would have hoped. Instead, it opened up the possibility of examining other candidate field equations and applying to them the extensive network of conclusions that were first developed for the *Entwurf* theory. The linking of this new possibility with the previously discovered weak points of the *Entwurf* theory moved Einstein, after a contemplative phase, to give up his attempts to strengthen the *Entwurf* theory and to return instead to a new exploratory phase.

A careful analysis discloses that practically all the technical problems that had arisen in Einstein's pursuit of the mathematical strategy in the Zurich Notebook could be solved in the following two years. This was a direct result of his preoccupation with themes that were tied to the *Entwurf* theory. Thus, although this theory had to be abandoned, it did play an essential role in the evolution of the general theory of relativity.

Einstein's results for the *Entwurf* theory paradoxically not so much solidified it but rather constituted tools that helped him remove the stumbling blocks that had earlier prevented him from accepting gravitational field equations based on the Riemann tensor.

### THE FINAL EFFORT: NOVEMBER 1915

In the fall of 1915, Einstein embarked on a new effort that culminated in November with the submission of four articles to the Royal Prussian Academy of Sciences. He began with a paper that brought him back to the mathematical road, which he had left three years earlier, convinced that he had now found the definitive solution. Only seven days later, he published an addendum offering a new interpretation of the same theory coupled with the provocative, though misguided, claim that all matter is of electromagnetic origin. One week later he presented a strong empirical argument in favor of the new theory, demonstrating its agreement with the observed precession of Mercury's perihelion. Finally, another week later, he revised his theory once more, removing the last stumbling block toward a generally covariant theory, or as the historian of science Michel Janssen put it, untying the knot that Einstein himself had created with the *Entwurf* theory.[36] When Einstein sent his academy papers to Arnold Sommerfeld in December 1915, he urged him to read them carefully despite the fact "that as you are reading, the final part of the battle for the field equations unfolds right in front of your eyes."

**NOVEMBER 4TH** In the first of these articles "On the General Theory of Relativity,"[37] submitted on November 4th, Einstein explained his change of perspective and his renewed search for a covariant field equation: "My efforts in recent years were directed towards basing a general theory of relativity, also for non-uniform motion, upon the supposition of relativity. I believed indeed to have found the only law of gravitation that complies with a reasonably formulated postulate of general relativity. . . . I lost trust in the field equations I have derived. . . ." He then recalled regretfully: "I arrived at the demand of general covariance, a demand from which I parted, though with a heavy heart, three years ago when I worked with my friend Grossmann."

It turned out that a slight adjustment of the expression for the gravitational field itself was actually sufficient to turn the mathematical apparatus developed for the *Entwurf* theory to obtain the theory of November 4th, the so-called November theory. The new expression for the gravitational field was the Christoffel symbol, which is a combination of derivatives of the components of the metric tensor. It is constructed in a more complicated way than the corresponding expression that represented the gravitational field in the case of the *Entwurf* theory. In this paper Einstein referred to the former version of the gravitational field as a "fatal prejudice," whereas a few weeks later, in a letter to Arnold Sommerfeld, he described the identification of the gravitational field with the Christoffel symbol as "the key to the solution."

The Christoffel symbol is a familiar quantity from the absolute differential calculus, and it appears naturally in discussions of the main objects of interest in this calculus: the Riemann and Ricci tensors. If the Christoffel symbol is interpreted as the proper mathematical expression for the gravitational field, an entirely new perspective opens up. Essentially, Einstein needed only to insert the new expression for the gravitational field into the mathematical formalism of the *Entwurf* theory to obtain a new field equation. This equation was strikingly similar to one of the candidates that he had already analyzed in the Zurich Notebook.

The November theory did not implement the general relativity principle completely, because there was still a slight constraint on the admissible coordinate systems, which was implied by the conservation principle. This principle thus necessitated, just as it did in the case of the *Entwurf* theory, a coordinate restriction. However, Einstein's feeling of triumph on publishing this theory was fully justified. By adapting the physical arguments used to derive the *Entwurf* theory, he succeeded in reaching a field equation that could be derived by the mathematical strategy, that is, by starting from the Riemann tensor and thus assuring a broad covariance. Thus the two strategies, the physical and the mathematical, had essentially converged; the pieces of the puzzle had assembled in a surprising new way. In retrospect everything depended on the reinterpretation of the gravitational field as expressed by the Christoffel symbol.

A few unanswered questions still remained. In particular, the physical meaning of the coordinate restriction that followed from the conservation principle was not clear. As minor as this restriction may have been, it was still not amenable to an obvious physical interpretation. So the tension between the mathematical formalism and the physical meaning of the new theory was not entirely dissipated.

**November 11th** To resolve this tension, on November 11th Einstein submitted an addendum[38] to the November 4th paper, which had just appeared in print. In the introduction he wrote: "I now want to show here that an even more concise and logical structure of the theory can be achieved by introducing an admittedly bold additional hypothesis on the structure of matter." Einstein thus proposed a new interpretation of the formalism of the November theory that led him to revive another candidate from the era of the Zurich Notebook: the Ricci tensor. If the assumption is made that the only fields occurring as sources of gravitation are electromagnetic fields, so that ultimately all matter can be reduced to the latter, then a field equation based on the Ricci tensor can be formulated without imposing any further coordinate restriction. With the introduction of this bold assumption, the conservation principle no longer implies any constraint on the admissible coordinate systems but instead becomes a stipulation on admissible sources of the gravitational field. In other words, by temporarily considering a speculative hypothesis

November 1915: Einstein could finally free himself from any restrictions on the choice of coordinates.

on the nature of matter, Einstein moved one step closer to his ultimate goal of a generally covariant theory of gravitation.

**NOVEMBER 18TH** On November 18th, Einstein submitted another paper to the Prussian academy, "Explanation of Perihelion Motion of Mercury from the General Theory of Relativity."[39] This is the only one of the November papers that he also presented orally to the academy—apparently hoping to enlist further support for the verification of his theory from the astronomers. Einstein was so confident about his new theory that he was now willing to make the effort to apply the method he had worked out with Besso—in the context of the *Entwurf* theory—to calculate the perihelion precession, and he obtained the correct result. When Einstein saw this result, he was so excited that, as he told one of his colleagues, he had heart palpitations. One day after submitting this paper, David Hilbert wrote to him: "Many thanks for your postcard and congratulations on conquering perihelion motion. If I could calculate as rapidly as you, in my equations the electron would correspondingly have to capitulate, and simultaneously the hydrogen atom would have to produce its note of apology about why it does not radiate." Einstein accepted the compliment but omitted to tell Hilbert that there was no need to start from scratch—the only requirement was a slight modification of his earlier calculations with Besso.

In the context of the November theory, Einstein made another important discovery. For the first time he realized that the coordinate conditions that followed from the correspondence principle had an entirely different meaning from the coordinate restrictions that followed from the conservation principle. To satisfy the Newtonian limit, it was naturally legitimate to select, from among the multitude of allowable coordinates, the coordinate system that naturally described a system at rest, corresponding to the conventional choice in Newton's gravitational theory. In the modern understanding, the choice of a particular coordinate system is thus only a matter of convenience rather than being imposed by the theory itself. In principle, this was the case also for Einstein's November theory, although the conservation principle still imposed some minor coordinate restrictions.

**NOVEMBER 25TH** In the last of the four papers,[40] submitted on November 25th, Einstein carried out the final crucial step following the inner logic of his research program. According to this logic, a theory that fully realizes the general principle of relativity should not be restricted by coordinate restrictions due to the conservation principle.

This goal motivated Einstein to overcome the remaining central tension in the relation between mathematical formalism and physical interpretation. The tension expressed itself either in a physically meaningless coordinate restriction (in the case of the theory of November 4th) or in a speculative hypothesis about the structure of matter (in the case of the theory based on the Ricci tensor, presented on November 11th). Both assumptions proved to be superfluous in the final version of the theory. All that was required to achieve this final version was to change the way in which the sources of the gravitational field were inserted on the right-hand side of the gravitational field equation. If the trace of the energy-momentum tensor, that is, the sum of its diagonal components, is appropriately added to the source term on the right-hand side of the field equation, then all the additional conditions become superfluous. In particular, the conservation principle is also satisfied as an automatic consequence of the modified field equation. Alternatively, the modified expression now known as the Einstein tensor can be used on the left-hand side instead of the Ricci tensor; however, Einstein came to this final modification of his theory starting from the right-hand side of the field equation. This modification became possible only after Einstein had learned—in the context of the calculation of the perihelion motion of Mercury—how to properly interpret the Newtonian limit. Contrary to what he had believed while working on the Zurich Notebook, Einstein found that the additional term on the right-hand side of the field equation indeed did not disturb its compatibility with the Newtonian limit.

In his later writings Einstein frequently emphasized that the new solution to the problem of gravitation is a natural consequence of the mathematical theory centered on the Riemann tensor, from which the candidates of the mathematical strategy can be obtained. So he himself described the breakthrough of late 1915 not as the result of a convergence of physical and mathematical strategies but as an exclusive success of the latter. Even in his first November paper, Einstein was fascinated by the power of mathematical formalism to lead to the correct theory: "Nobody who really grasped it can escape from its charm, because it signifies a real triumph of the general differential calculus as founded by GAUSS, RIEMANN, CHRISTOFFEL, RICCI AND LEVI-CIVITA." (The names are capitalized in the original publication.)

### THE FINAL RACE: ALBERT EINSTEIN VERSUS DAVID HILBERT

While Einstein was working on the final stages of his theory, a parallel effort was taking place in Göttingen, where David Hilbert, the acknowledged leading mathematician of his generation, presented his famous paper "On the Foundations of Physics (First Contribution)." The published version of this talk to the Göttingen academy contains the correct field equation of the theory of general relativity and is dated 20 November 1915, that is, five days before the submission of Einstein's final paper. Although Hilbert's paper was not published until 1916, it is often claimed that he should be given priority over Einstein for formulating the field equation. At first, Einstein was also concerned that Hilbert might claim priority for himself. This was a source of bitter, though brief, contention between these two personal and professional friends.[41]

Hilbert and Einstein—
who got there first?

The consensus on Hilbert's priority underwent a dramatic reversal when galley proofs of Hilbert's academy talk, carrying the stamp "6 December 1915"—thus dating them after Einstein's conclusive paper—were uncovered in his archive.[42] The theory developed in Hilbert's proofs differs distinctly in some important respects from the published version, so he must have considerably revised the version documented in the proofs before publication. It turned out that the conceptual basis of the proof version of Hilbert's theory is in many respects more similar to Einstein's *Entwurf* theory than to the final version of the theory of general relativity. Against this background, it now appears unlikely that Hilbert had the key for Einstein's problems at hand or that his contribution represented a triumph of a mathematical strategy that needed no interaction with a physical strategy.

In his paper, Hilbert graciously acknowledged Einstein's primacy: "The differential equations of gravitation that result here are, as it seems to me, in agreement with the magnificent theory of general relativity established by Einstein."[43]

## THE 1916 MANUSCRIPT: NOT THE END OF THE STORY

In the fall of 1914, Einstein's confidence in the *Entwurf* theory had grown to the point that he was encouraged to write an extensive review article, "The Formal Foundation of the General Theory of Relativity," published in the meeting reports of the Royal Prussian Academy of Sciences. Referring to the 1915 academy papers, Einstein later wrote to Sommerfeld: "unfortunately I have immortalized my final errors in this battle in the academy papers."[44] Referring to his 1914 review article in a letter to his friend Ehrenfest, Einstein was even more ironic: "It's convenient with that fellow Einstein, every year he retracts what he wrote the year before."[45]

After November 25, 1915, Einstein was ready to summarize his general theory of relativity in the manuscript reproduced in this book. The first two parts of this manuscript closely follow his 1914 summary article based on the *Entwurf* theory. The introduction of the field equation and the discussion of the energy-momentum conservation in the manuscript are very different from the corresponding chapter in the 1914 article. In the former they closely follow a letter to Paul Ehrenfest in which Einstein responded to questions and remarks by his close friend, who had served on many occasions as his sounding board. The last section contains the three basic predictions of the theory: the bending of light rays in the gravitational field of the sun, the gravitational redshift, and the precession of the perihelion of planet Mercury.

Appended to the manuscript on general relativity in this text is a five-page manuscript titled "Appendix: Formulation of the Theory on the Basis of a Variational Principle." Judging from the equation and paragraph numbers at the beginning of the manuscript, Einstein intended to include it in the core of the article. He then changed those numbers and added the title, indicating that he had decided to publish it as an appendix to the article. Ultimately he decided not to include it at all. About seven months later, he published a very similar article in the meeting reports of the Prussian Academy of Sciences, titled "Hamilton's Principle and the General Theory of Relativity."[46]

The addendum fills an important gap in the main body of Einstein's review paper. It establishes the connection between covariance and energy conservation without the restriction to unimodular coordinates prevailing in the main paper. Einstein had already made this connection, which would eventually culminate in Emmy Noether's famous theorems about the relation between invariance properties and conservation laws, in the *Entwurf* theory. But the formulation of his final theory using unimodular coordinates had impeded Einstein from carrying over this relation. This deficiency was removed in the addendum, as Einstein proudly related in correspondence with his friends and colleagues. Thus he wrote to Michele Besso: "You will soon receive a short paper of mine about the foundations of general relativity, in which it is shown how the requirement of relativity is connected with the energy principle. It is very amusing."[47]

The differences between the published version and the unpublished manuscript are striking, in particular, in Einstein's references to Hilbert and Lorentz. The addendum manuscript contains two brief footnotes, whereas in the published paper, these notes have a much more visible and prominent place in the opening paragraph, which sets the stage for the entire calculation. The elucidation of these differences sheds additional light on the final stages of Einstein's road to his theory of general relativity.

The end (?) .

The end? The show must go on!

These introductory notes have reconstructed the complex process through which Einstein's heuristics led to the formulation of the general theory of relativity in the year 1915. In particular, the interaction between Einstein's heuristics and his intermediate mathematical results played a key role. These concrete results acquired a new physical interpretation, thereby altering the heuristics. However, this interaction did not end with the formulation of the field equation in the concluding paper of November 1915. The tensions between Einstein's heuristics and the implications of the new theory also characterized its further evolution until at least 1930, and in some respects this process continues even today.

## NOTES

1. See John Stachel (ed.), *Einstein's Miraculous Year: Five Papers that Changed the Face of Physics* (Princeton, NJ: Princeton University Press, 2005).
2. See the prologue to the article by John Stachel, "The First Two Acts," in *The Genesis of General Relativity*, vol. 1, pp. 81ff.
3. See Philipp Frank (ed.), *Einstein: His Life and Times* (New York: Da Capo, 1989).
4. Albert Einstein, "On the Relativity Principle and the Conclusions Drawn from It" (1907), in CPAE vol. 2, Doc. 47, 252–311.
5. Max Abraham, "Zur Theorie der Gravitation," *Physikalische Zeitschrift* 13 (1912): 1–4. English translation in *The Genesis of General Relativity*, vol. 3, pp. 331–339.
6. Einstein to Michele Besso, 26 March 1912, in CPAE vol. 5, Doc. 377, pp. 276–279.
7. The full introduction is reprinted in CPAE vol. 6, Doc. 42, p. 418.
8. The "Zurich Notebook" is described and analyzed in *Einstein's Zurich Notebook: Commentary and Essays*, vol. 2 of *The Genesis of General Relativity*.
9. Albert Einstein and Marcel Grossmann, *Outline of a Generalized Theory of Relativity and of a Theory of Gravitation* (1913), in CPAE vol. 4, Doc. 13, pp. 151–188.
10. The "Einstein-Besso Manuscript" is described and analyzed in "What Did Einstein Know and When Did He Know It? A Besso Memo Dated August 1913," in *The Genesis of General Relativity*, vol. 2, pp. 785–837. See also John Earman and Michel Janssen, "Einstein's Explanation of the Motion of Mercury's Perihelion," in *The Attraction of Gravitation*, vol. 5 of *Einstein Studies*, ed. J. Earman, M. Janssen, and J. D. Norton (Boston: Birkhäuser, 1993), 129–172.
11. Einstein to H. A. Lorentz, 14 August 1913, in CPAE vol. 5, Doc. 467, pp. 349–351.
12. Einstein to Heinrich Zangger, 27 June 1914, in CPAE vol. 8, Doc. 16a; reprinted in vol. 10, Doc. 349a, pp. 11–12.
13. See Hanoch Gutfreund, "Einstein's Jewish Identity," in *Einstein for the 21st Century: His Legacy in Science, Art, and Modern Culture* (Princeton, NJ: Princeton University Press, 2008); and John Stachel, "Einstein's Jewish Identity," in *Einstein from 'B' to 'Z'* (Basel: Birkhäuser, 2002). For Einstein's thoughts on politics, see David E. Rowe and Robert Schulmann (eds.), *Einstein on Politics: His Private Thoughts and Public Stands on Nationalism, Zionism, War, Peace, and the Bomb* (Princeton, NJ: Princeton University Press, 2007).
14. Albert Einstein, "The Formal Foundation of the General Theory of Relativity" (1914), in CPAE, vol. 6, Doc. 9, pp. 30–84.
15. Gerald Holton, "Who Was Einstein? Why Is He Still So Alive?" in *Einstein for the 21st Century*.
16. See Scott Walter, "Breaking in the 4-Vectors: the Four-Dimensional Movement in Gravitation, 1905–1910," in *The Genesis of General Relativity*, vol. 3, pp. 194–252.
17. For Mach's critique of classical mechanics, see Ernst Mach, *The Science of Mechanics: A Critical and Historical Account of Its Development* (LaSalle, Ill.: Open Court Publ., 1960).
18. See Jürgen Renn, "Classical Physics in Disarray," and John Stachel, "The First Two Acts," both in *The Genesis of General Relativity*, vol. 1, pp. 21–80 and pp. 81–111, respectively.
19. Albert Einstein, *Autobiographical Notes*, ed. P. A. Schilpp (La Salle IL: Open Court, [1949] 1979), 61.
20. See John Stachel, "Albert Einstein: A Man for the Millennium," in *International Conference on the Albert Einstein's Century, 11–22 July 2005*, ed. J. Alimi and A. Fufza (Paris, France: Melville: American Institute of Physics Press, 2006), 211–244; and "Einstein's Intuition and the Post-Newtonian Approximation,"

in *Proceedings of the Conference Topics in Mathematical Physics, General Relativity and Cosmology on the Occasion of the 75th Birthday of Professor Jerzy F. Plebanski* (Mexico City: World Scientific, 2002), 453–467.

21. See Michel Janssen, "The Twins and the Bucket: How Einstein Made Gravity Rather than Motion Relative in General Relativity," *Studies in History and Philosophy of Modern Physics* 43(2012): 159–175.
22. "der glücklichste Gedanke meines Lebens," CPAE vol. 7, Doc. 31, p. 136 [p. 21].
23. See the discussion in John Stachel, "The Rigidly Rotating Disk as the 'Missing Link' in the History of General Relativity," in *Einstein and the History of General Relativity*, ed. D. Howard and J. Stachel, vol. 1 of *Einstein Studies* (Boston: Birkhäuser, 1989), 48–62. See also the discussion in Michel Janssen "'No success like failure . . .': Einstein's Quest for General Relativity, 1907–1920," in *The Cambridge Companion to Einstein*, ed. M. Janssen and C. Lehner (Cambridge: Cambridge University Press, 2014), 167–227 [p. 181].
24. Albert Einstein, "The Speed of Light and the Statics of the Gravitational Field" (1912), reprinted in CPAE vol. 4, Doc. 3, pp. 95–106.
25. See Jürgen Renn and Tilman Sauer, "Pathways out of Classical Physics," in *The Genesis of General Relativity*, vol. 1, pp. 133–312.
26. See note 1.
27. Einstein to H. A. Lorentz, 16 August 1913, in CPAE vol. 5, Doc. 470, pp. 352–353.
28. Einstein to Ernst Mach, second half of December 1913, in CPAE vol. 5, Doc. 495, pp. 370–371.
29. See John Stachel, "The Hole Argument and Some Physical and Philosophical Implications," *Living Reviews in Relativity* 17 (2014): 1–66. http://www.livingreviews.org/lrr-2014–1.
30. Einstein to Ludwig Hopf, 2 November 1913, CPAE vol. 5, Doc. 480, pp. 358–359.
31. Einstein to Michele Besso from ca. 10 March 1914, in CPAE vol. 5, Doc. 514, pp. 381–382.
32. See Albert Einstein and Marcel Grossmann, "Covariance Properties of the Field Equations of the Theory of Gravitation Based on the Generalized Theory of Relativity" (1914), in CPAE vol. 6, Doc. 2, pp. 6–15.
33. Einstein to Conrad Habicht, 24 December 1907, in CPAE vol. 5, Doc. 69, p. 47.
34. The Einstein-Besso manuscript (see note 10).
35. See Michel Janssen, "Rotation as the Nemesis of Einstein's Entwurf Theory," in *The Expanding Worlds of General Relativity,* ed. H. Goenner, J. Renn, J. Ritter and T. Sauer, vol. 7 of *Einstein Studies* (Boston: Birkhäuser, 1999). For a more concise and updated version, see also Michel Janssen "'No success like failure . . .'," 167–227.
36. For a more detailed technical version of the account of the developments of November 1915, see Michel Janssen and Jürgen Renn, "Untying the Knot: How Einstein Found His Way Back to Field Equations Discarded in the Zurich Notebook," in *The Genesis of General Relativity*, vol. 2, pp. 839–925.
37. Albert Einstein, "On the General Theory of Relativity" (1915), in CPAE vol. 6, Doc. 21, pp. 98–107.
38. Albert Einstein, "On the General Theory of Relativity (Addendum)" (1915), in CPAE vol. 6, Doc. 22, pp. 108–110.
39. Albert Einstein, "Explanation of the Perihelion Motion of Mercury from the General Theory of Relativity" (1915), in CPAE vol. 6, Doc. 24, pp. 112–116.
40. Albert Einstein, "The Field Equations of Gravitation" (1915), in CPAE 6, Doc. 25, pp. 117–120.
41. Einstein to Heinrich Zangger, 26 November 1915, in CPAE vol. 8, Doc. 152, pp. 150–151.
42. See Leo Corry, Jürgen Renn, and John Stachel, "Belated Decision in the Hilbert-Einstein Priority Dispute," *Science* 278 (1997): 1270–1273.
43. David Hilbert, "Die Grundlagen der Physik (Erste Mitteilung)," *Königliche Gesellschaft der Wissenschaften zu Göttingen. Mathematisch-Physikalische Klasse. Nachrichten* (1915): 395–407.
44. Einstein to Arnold Sommerfeld, 28 November 1915, CPAE vol. 8, Doc. 153, pp. 152–153.
45. Einstein to Paul Ehrenfest, 26 December 1915, CPAE vol. 8, Doc. 173, pp. 167–168.
46. *Hamilton's Principle and the General Theory of Relativity* (referred to as the "October paper"), in CPAE vol. 6, Doc. 41, pp. 240–245.
47. Einstein to Michele Besso, 31 October 1916, CPAE vol. 8, Doc. 270, p. 257–259.

# THE ANNOTATED MANUSCRIPT

THE ANNOTATIONS FACING THE REPRODUCTIONS OF THE ORIGINAL MANUSCRIPT PAGES are set in three different type styles to differentiate their content. Text type 1 (plain text) refers to the content of the specific manuscript page, whereas text type 2 (set off by rule) refers to contextual background material. The boxed text explains a specific idea or concept. The bibliographic notes appear at the end of the chapter. The page numbers of the manuscript are given in square brackets in the header, next to the actual page numbers of this book. This numeration in square brackets is also reflected in the cross referencing of this section.

Die Grundlage der allgemeinen Relativitätstheorie.

A. Prinzipielle Erwägungen zum Postulat der Relativität.

§1. Die spezielle Relativitätstheorie.

Die im Nachfolgenden dargelegte Theorie bildet die denkbar weitgehendste Verallgemeinerung der heute allgemein als „Relativitätstheorie" bezeichneten Theorie; die letztere nenne ich im folgenden zur Unterscheidung von der ersteren „spezielle Relativitätstheorie" und setze sie als bekannt voraus. Diese Verallgemeinerung der Relativitätstheorie wurde sehr erleichtert durch die Gestalt, welche der speziellen Relativitätstheorie durch Minkowski gegeben wurde, welcher Mathematiker zuerst die formale Gleichwertigkeit der räumlichen Koordinaten und der Zeitkoordinate klar erkannte und für den Aufbau der Theorie nutzbar machte. Die für die allgemeine Relativitätstheorie nötigen mathematischen Hilfsmittel lagen fertig bereit in dem „absoluten Differentialkalkül", welcher auf den Forschungen von Gauss, Riemann und Christoffel über nichteuklidische Mannigfaltigkeiten ruht und von Ricci und Levi-Civita in ein System gebracht und bereits auf Probleme der theoretischen Physik angewendet wurde. Ich habe im Abschnitt B der vorliegenden Abhandlung alle für uns nötigen, bei dem Physiker nicht als bekannt vorauszusetzenden mathematischen Hilfsmittel in möglichst einfacher und durchsichtiger Weise entwickelt, sodass ein Studium mathematischer Literatur für das Verständnis der vorliegenden Abhandlung nicht erforderlich ist. Endlich sei an dieser Stelle dankbar meines Freundes, des Mathematikers Grossmann gedacht, der mir durch seine Hilfe nicht nur das Studium der einschlägigen mathematischen Literatur ersparte, sondern mich auch beim Suchen nach den Feldgleichungen der Gravitation unterstützte. —

A. Prinzipielle Erwägungen zum Postulat der Relativität.

§1. Bemerkungen zu der speziellen Relativitätstheorie.

Der speziellen Relativitätstheorie liegt folgendes Postulat zugrunde, welchem auch durch die Galilei-Newton'sche Mechanik Genüge geleistet wird: Wird ein Koordinatensystem K so gewählt, dass in bezug auf dasselbe die physikalischen Gesetze in ihrer einfachsten Form gelten, so gelten dieselben Gesetze auch in bezug auf jedes andere Koordinatensystem K', das relativ zu K in gleichförmiger Translationsbewegung begriffen ist. Dies Postulat nennen wir R, „spezielles Relativitätsprinzip." Durch das Wort „speziell" soll angedeutet werden, dass das Prinzip auf den

## Why did Einstein go beyond special relativity?

Why in 1905 did Einstein formulate the theory of special relativity in the first place? Its main achievement was an extension of the Galilean-Newtonian relativity principle, which stipulates that the laws of mechanics are the same in all inertial frames of reference that move with constant velocity with respect to each other. Einstein extended this principle to all laws of physics. The classical relativity principle can be described by the mental model of a train (with blocked windows) moving at constant velocity. There is no *mechanical* measurement that passengers on that train can perform that will tell them if they are at rest or moving with respect to the platform.

Can this relativity principle be extended to all physical phenomena, including electromagnetic phenomena such as light? According to the then-prevailing interpretation of these phenomena, based on Maxwell's equations, this seemed hardly possible. Light was known to be a wave phenomenon, and such phenomena require the existence of a medium in which to propagate. In the case of electromagnetism this medium was called the "ether," but it turned out to be impossible to detect it by experiments. The ether was assumed to be immobile and to constitute a preferred system of reference in which the velocity of light, appearing explicitly in Maxwell's equations, is a constant. Einstein made the bold assumption, which is incompatible with classical physics, that this constancy holds in all inertial frames and that the ether does not exist, thus extending the relativity principle to all physical phenomena. If the velocity of light were not constant, the laws of electromagnetism would be different in different inertial frames.

In a popular account of the special and general theory of relativity, published in 1917, Einstein stated: "Since the introduction of the special principle of relativity has been justified, every intellect which strives after generalization must feel the temptation to venture the step towards the general principle of relativity." While the very few "intellects" who were actually tempted to venture that step worked at the margins of physics and remained unsuccessful, Einstein was the only one who had persistently followed this intuition since 1907.

The present manuscript summarizes Einstein's successful conclusion of this effort to go beyond special relativity. On the first page he shifts the intended titles of part A and section 1 farther down the page to include a few introductory remarks in which he mentions the names of people who played an essential role in his discovery of general relativity:

- Hermann Minkowski, who developed a geometric formulation of the special theory of relativity in terms of a four-dimensional spacetime, which became a natural starting point in the transition from the special to the general theory;
- Carl Friedrich Gauss, the founding father of geometry on curved surfaces;
- Bernhard Riemann, Elwin Bruno Christoffel, Gregorio Ricci-Curbastro, and Tullio Levi-Civita, the mathematicians who extended the work of Gauss to higher dimensions and developed the necessary mathematical concepts and methods;
- Marcel Grossmann, to whom Einstein owed special gratitude for introducing him to the mathematical tools and for working with him during the early stages of the development of general relativity.

Fall beschränkt ist, dass $K'$ eine gleichförmige Translationsbewegung gegen $K$ ausführt, dass sich aber die Gleichwertigkeit von $K'$ und $K$ nicht auf den Fall ungleichförmiger Bewegung von $K'$ gegen $K$ erstreckt.

Die spezielle Relativitätstheorie weicht also von der klassischen Mechanik nicht durch das Relativitätspostulat, sondern allein durch das Postulat der Konstanz der Vakuum-Lichtgeschwindigkeit ab, aus welchem im Verein mit dem speziellen Relativitätsprinzip die Relativität der Gleichzeitigkeit sowie jene die Lorentztransformation und die mit dieser verknüpften Gesetze über das Verhalten bewegter starrer Körper und Uhren in bekannter Weise folgen.

Die Modifikation, welche die Lehre von Raum und Zeit durch die spezielle Relativitätstheorie erfahren hat, ist zwar eine tiefgehende. Aber ein wichtiger Punkt blieb unangetastet. Auch gemäss der speziellen Relativitätstheorie sind nämlich die Sätze der Geometrie unmittelbar als die Gesetze über die möglichen relativen Lagen (ruhender) fester Körper aufzufassen zu deuten, allgemeiner die Sätze der Kinematik als Sätze, welche das Verhalten von Messkörpern und Uhren beschreiben. Zwei hervorgehobenen materiellen Punkte eines ruhenden (starren) Körpers entspricht hierbei stets eine Strecke von ganz bestimmter Länge, unabhängig von Ort und Orientierung des Körpers, sowie von der Zeit. Zwei hervorgehobenen Zeiger- stellungen einer relativ zum Bezugssystem (berechtigten) ruhenden Uhr entspricht stets eine Zeitstrecke von bestimmter Länge, unabhängig von Ort und Zeit. Es wird sich bald zeigen, dass die allgemeine Relativi- tätstheorie an dieser einfachen physikalischen Deutung von Raum und Zeit nicht festhalten kann.

### §2. Über die Gründe, welche eine Erweiterung des Relativitätspostulates nahelegen.

Der klassischen Mechanik und nicht minder der speziellen Relativitätstheorie haftet ein erkenntnistheoretischer Mangel an, der wohl zum ersten Male vielleicht insbesondere von E. Mach hier hervorgehoben wurde. Wir erläutern ihn an folgendem Beispiel. Zwei flüssige Körper von gleicher Grösse und Art schweben frei im Raume in so grosser Entfernung voneinander (und von allen übrigen Massen), dass nur diejenigen Gravitationskräfte berücksichtigt werden müssen, welche die Teile eines dieser Körper aufeinander ausüben. Die Entfernung der Körper voneinander sei unveränderlich, die Verbindungslinie aber Schwerpunkte zeigen Relative Bewegungen der Teile eines Körpers gegeneinander sollen nicht auftreten. Aber jede Masse soll als Ganzes — von einem relativ zu der andern Masse ruhenden Beobachter aus beurteilt — um die Verbindungs- linie der Massen mit konstanter Winkelgeschwindigkeit rotieren ( es ist dies eine konstatierbare Relativbewegung beider Massen). Nun denken

## What was wrong with the classical notions of space and time?

The classical notions of space and time were based not only on practical experiences with rulers and clocks, but they also worked well as a foundation of classical mechanics and astronomy—indeed, so well that hardly anyone doubted they would ever change. Yet, there were challenges.

Einstein's extension in 1905 of the classical relativity principle to include electromagnetic phenomena was based on a reinterpretation of electromagnetism as developed by James Clerk Maxwell, Heinrich Hertz, and Hendrik Antoon Lorentz. From electromagnetism, Einstein extracted the principle that the velocity of light must always remain constant. Together, the relativity principle and the constancy of the velocity of light imply fundamental modifications in the conception of space and time. Time and space lose their absolute meaning in the sense that such global concepts as a particular slicing of space and time are recognized to be of a merely conventional character, but this is not what is physically relevant. What is relevant is that, in contrast with classical physics, local time in the sense of time measured by a clock depends on the path between two events. The special theory of relativity implies far-reaching modifications in the notion of space and time, but one aspect of space and time measurements remains unaffected. In classical physics and in special relativity, the laws of geometry can be interpreted directly as laws concerning the possible placements of a rigid body at rest. The same applies to the time interval between two selected positions of the hands of a clock at rest with respect to a given reference frame. Einstein alerts the reader to the fact that this simple physical interpretation of space and time will have to be abandoned in the transition to general relativity.

Einstein begins his exposition of the transition from the special to the general theory of relativity by pointing to a basic epistemological defect in classical Newtonian mechanics, which was first emphasized by Ernst Mach. He even claims that this deficit holds for both classical mechanics and special relativity, without realizing that special relativity had in some sense already solved this problem. Einstein demonstrates the problem by the example of two bodies, one rotating and the other not. This is his version of the celebrated Newton's "two buckets experiment" (also next page).

Newtonian mechanics is based on the notion that space is absolute by its nature—without relation to extraneous objects, always the same and immovable—and that there is a distinction between relative and absolute motion. Isaac Newton explains this distinction using his bucket experiment. One of two identical buckets filled with water is rotating. The parabolic surface of water in the rotating bucket is caused by the centrifugal forces, which are present only in this bucket. According to Newton, the rotation occurs with respect to absolute space, and the presence of centrifugal forces distinguishes between absolute and relative motion. Mach rejected the notion of absolute space and absolute motion, suggesting that the shape of the water surface in the rotating bucket can be explained as being produced by rotation with respect to the rest of the matter in the universe. The same effect would occur if the bucket were at rest and the universe were rotating. Einstein fully embraced Mach's way of thinking and later referred to it as "Mach's principle." According to Einstein "Mach clearly recognized the weak points of classical mechanics and thus came close to demand a general theory of relativity" (in Mach's obituary, 1916).

(3)

wir uns die Oberflächen beider Körper ($S_1$ und $S_2$) mit Hilfe (relativ ruhender) Maßstäbe ausgemessen; es ergebe sich, dass die Oberfläche von $S_1$ eine Kugel, die von $S_2$ ein Rotationsellipsoid sei.

Wir fragen nun: Aus welchem Grunde verhalten sich die Körper $S_1$ und $S_2$ verschieden? Eine Antwort auf diese Frage kann nur dann als befriedigend erkenntnistheoretisch anerkannt erklärt werden, wenn die als Grund angegebene Sache eine beobachtbare Thatsache ist; denn das Kausalitätsgesetz hat nur dann den Sinn einer Aussage über die Erfahrungswelt, wenn als Ursachen und Wirkungen letzten Endes nur beobachtbare Thatsachen auftreten.

Die Newton'sche Mechanik gibt auf diese Frage keine befriedigende Antwort. Sie sagt nämlich folgendes. Die Gesetze der Mechanik gelten wohl für einen Raum $R_1$, gegen welchen der Körper $S_1$ in Ruhe ist, nicht aber gegenüber einem Raume $R_2$, gegen welchen $S_2$ in Ruhe ist. Der berechtigte Galileische Raum $R_1$, der hiebei eingeführt wird ist aber eine bloss fingierte Ursache, keine beobachtbare Sache. Es ist also klar, dass die Newton'sche Mechanik der Forderung der Kausalität in dem betrachteten Falle nicht wirklich, sondern nur scheinbar Genüge leistet, indem sie die bloss fingierte Ursache $R_1$ für das beobachtbare verschiedene Verhalten der Körper $S_1$ und $S_2$ verantwortlich macht.

Eine befriedigende Antwort auf die oben aufgeworfene Frage kann nur so lauten. Das aus $S_1$ und $S_2$ bestehende physikalische System zeigt für sich allein keine denkbare Ursache, auf welche das verschiedene Verhalten von $S_1$ und $S_2$ zurückgeführt werden könnte. Die Ursache muss also ausserhalb dieses Systems liegen. Man gelangt zu der Auffassung, dass die allgemeinen Bewegungsgesetze, welche im Speziellen die Gestalten von $S_1$ und $S_2$ bestimmen, derart sein müssen, dass das mechanische Verhalten von $S_1$ und $S_2$ ganz wesentlich durch ferne Massen mitbedingt werden muss, welche wir nicht zu dem betrachteten System gerechnet hatten. Diese fernen Massen und deren Relativbewegungen gegen die betrachteten Körper sind dann als prinzipiell beobachtbare Ursachen für das verschiedene Verhalten unserer betrachteten Körper anzusehen; sie übernehmen die Rolle der fingierten Ursache $R_1$. Von allen denkbaren, relativ zueinander bewegten Räumen $R_1$, $R_2$ etc. darf a priori keiner als bevorzugt angesehen werden, wenn nicht der dargelegte erkenntnistheoretische Einwand wieder aufleben soll. Die Gesetze der Physik müssen so beschaffen sein, dass sie bezüglich auf beliebig bewegte Bezugsysteme gelten. Wir gelangen also auf diesem Wege zu einer Erweiterung des Relativitätspostulates.

× Eine derartige erkenntnistheoretisch befriedigende Antwort, kann natürlich immer noch physikalisch unzutreffend sein, falls sie mit anderen Erfahrungen im Widerspruch ist.

## Why did Einstein see difficulties that others ignored?

At the beginning of the twentieth century, classical physics was at its apex. It successfully explained many phenomena. When new experiments revealed new insights, such as the discovery of new forms of radiation, they were not considered as challenges to its fundamental concepts such as space and time. Why did Einstein question these foundations?

Einstein was not primarily a philosopher but, like his colleagues, was interested in physical explanations. Yet, he appreciated epistemological thinking in the search for basic principles and physical theories. In his formulation of special and general relativity, he thought on several levels: that of the concrete phenomena to be explained, that of the available mathematical and theoretical tools, and that of the physical concepts employed. He was aware that these concepts were tentative human constructs and not given a priori and hence could be modified.

According to his own testimony, Einstein was deeply influenced by the philosopher David Hume and the physicist-philosopher Ernst Mach: "Nobody can deny that epistemologists paved the role for progress; and for myself, I know at least that Hume and Mach have helped me a lot, both directly and indirectly." He returned to this point also much later when he wrote, at age 67, his *Autobiographical Notes*, commenting on how difficult it was to get rid of the belief in the absolute nature of simultaneity: "To recognize clearly this axiom and its arbitrary character already implies the essentials of the solution of the problem. The type of critical reasoning required for the discovery of this central point was furthered, in my case especially, by the reading of David Hume's and Ernst Mach's philosophical writings."

During his years in Bern (1902–1908) Einstein formed a reading club with two friends, Maurice Solovine and Conrad Habicht, which they called "Olympia Academy." They read, among others, Mach's *The Analysis of Sensations and the Relation of the Physical to the Psychical*, and Hume's *A Treatise on Human Nature*. As a student Einstein also read Mach's *Mechanics*.

ETH-Bibliothek
Zürich, Bildarchiv

In the text, Einstein describes his own version of Newton's bucket argument. He refers to a system of two fluid bodies of the same size and nature, separated in space. One of them is rotating at constant angular velocity around the axis connecting the two bodies. An observer at rest, with respect to either body, sees the other one as rotating, yet the shape of only one of them is changed as a result of this rotation. Einstein cannot see any possible cause within the system of the two bodies for their different behaviors. He therefore concludes, following Mach, that the difference must be attributed to a cause outside the system, namely, the distant masses in the universe. None of the reference frames with respect to which one of the two bodies is at rest is privileged. This conclusion leads him to the general principle of relativity: The laws of physics must be of such a nature that they apply to systems of reference in any kind of motion.

(4)

Neben diesen schwerwiegenden erkenntnistheoretischen Argument spricht aber auch eine wohlbekannte physikalische Thatsache für eine Erweiterung der Relativitätstheorie. Es sei K ein Galileisches Bezugsystem, d. h. ein solches, relativ zu welchem (mindestens in dem betrachteten vierdimensionalen Gebiete) eine von anderen hinlänglich entfernte Masse sich gradlinig und gleichförmig bewegt. Es sei K' ein zweites Koordinatensystem, welches relativ zu K in gleichförmig beschleunigter Translationsbewegung sei. Relativ zu K' führt dann eine von anderen hinreichend getrennte Masse eine beschleunigte Bewegung aus, derart, dass deren Beschleunigung und Beschleunigungsrichtung von der stofflichen Zusammensetzung und ihrem physikalischen Zustande unabhängig ist.

Kann ein relativ zu K' ruhender Beobachter hieraus den Schluss ziehen, dass er sich auf einem wirklich beschleunigten Bezugssystem befindet? Diese Frage ist zu verneinen; denn das Verhalten frei beweglicher Massen relativ zu K' kann ebensogut auf folgende Weise gedeutet werden. Das Bezugssystem K' ist unbeschleunigt; in dem betrachteten zeiträumlichen Gebiete herrscht aber ein Gravitationsfeld, welches die beschleunigte Bewegung der Körper relativ zu K' erzeugt.

Diese Auffassung wird dadurch ermöglicht, dass uns die Erfahrung die Existenz eines Kraftfeldes (nämlich des Gravitationsfeldes) gelehrt hat, welches die merkwürdige Eigenschaft hat, allen Körpern dieselbe Beschleunigung zu erteilen.× Das Verhalten der Körper relativ zu K' ist dasselbe, wie es gegenüber Systemen sich der Erfahrung darbietet, die wir als „ruhende" bezw. als „berechtigte" Systeme anzusehen gewohnt sind. deshalb liegt es auch vom physikalischen Standpunkt nahe, anzunehmen, dass die Systeme K und K' beide mit demselben Recht als „ruhend" angesehen werden können, bezw. dass sie als Bezugssysteme für die physikalische Beschreibung der Vorgänge gleichberechtigt seien.

Aus diesen Erwägungen sieht man, dass die Durchführung der allgemeinen Relativitäts-Theorie zugleich zu einer Theorie der Gravitation führen muss. denn man kann ein Gravitationsfeld durch blosse Aenderung des Koordinatensystems „erzeugen". Ebenso sieht man unmittelbar, dass das Prinzip von der Konstanz der Lichtgeschwindigkeit eine Modifikation erfahren muss. Denn man erkennt leicht, dass die Bahn eines Lichtstrahles in bezug auf K' im Allgemeinen eine krumme sein muss, wenn sich das Licht in bezug auf K gradlinig fortpflanzt, und mit bestimmter, konstanter Geschwindigkeit fortpflanzt.

× Dass das Gravitationsfeld diese Eigenschaft mit grosser Genauigkeit besitzt, hat Eötvös experimentell bewiesen.

## What was Einstein's happiest thought and how did it come about?

In 1922, Einstein delivered a lecture at the University of Kyoto titled "How I Created the Theory of Relativity." There he recalled that "I was sitting in a chair in the patent office in Bern when all of a sudden a thought occurred to me that when a person falls freely he will not feel his own weight. I was stunned. This simple thought impressed me greatly. It led me to the theory of gravity." Later in life, Einstein referred to this revelation as the "happiest thought" of his life. Why was this idea so important for Einstein and what had motivated it? Let us first turn to the role of this idea in Einstein's work on a theory of gravitation.

On this page, Einstein notes that the gravitational field has the remarkable property of imparting the same acceleration to all bodies ("Galileo's principle"); gravitation and acceleration are to some extent interchangeable. Thus an observer in free fall does not sense the effect of gravitation and momentarily feels like an astronaut.

Similarly, an observer in a uniformly accelerated system of reference in outer space can interpret the motion of masses with respect to this system as if the system were at rest but exposed to a uniform static gravitational field. This is the core of Einstein's famous equivalence principle. It allowed him, to some extent, to study gravitational effects with the help of accelerated reference frames. Treating these reference frames with the help of the special theory of relativity, he was able to gain insights into relativistic properties of the gravitational field long before he had formatted a complete theory. Some of the most important heuristic hints toward the development of such a theory follow from the equivalence principle, for instance, the bending of a light in a gravitational field.

A beam of light that propagates in a straight line with respect to an inertial frame of reference will appear to be deflected in an accelerated frame of reference. From the equivalence principle, Einstein inferred that light must be bent in a gravitational field.

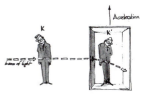

The observation of light bending during a solar eclipse in 1919 catapulted Einstein, almost overnight, to the status of world celebrity.

In classical mechanics, the mass of a material body plays a dual role. First, it determines the acceleration of the body caused by a given force. In this context it is called the *inertial mass*. Mass also determines the force on a body in a given gravitational field. In this context it is called the *gravitational mass*. In the footnote on this page, Einstein mentions the Hungarian physicist Lorand Eötvös, who proved to a great accuracy that these two masses are equal. From the perspective of classical physics, this equality appears as mere coincidence. When Einstein first attempted to formulate a relativistic theory of gravitation, it no longer seemed to hold. But then he realized that this equality automatically holds if one assumes the equivalence principle. This is why Einstein considered it his "happiest thought."

(5)

### § 3. Das Raum-Zeit-Kontinuum. Forderung der allgemeinen Kovarianz für die die allgemeinen Naturgesetze ausdrückenden Gleichungen.

In der klassischen Mechanik sowie in der speziellen Relativitätstheorie haben die Koordinaten des Raumes und der Zeit eine unmittelbare physikalische Bedeutung. Ein Punktereignis hat die $X_1$-Koordinate $x_1$, bedeutet: Die mittels eines starren Stabes nach den Regeln der euklidischen Geometrie ermittelte Projektion des Punktereignisses auf die $X_1$-Achse wird erhalten, indem man einen bestimmten Stab, den Einheitsmaßstab, $x_1$ mal vom Anfangspunkt des Koordinatenkörpers auf der (positiven) $X_1$-Achse abträgt. Ein Punkt hat die $X_4$-Koordinate $x_4 = t$, bedeutet: Eine relativ zum Koordinatensystem ruhend angeordnete, mit dem Punktereignis räumlich (praktisch) zusammenfallende Einheitsuhr, welche nach bestimmten Vorschriften gerichtet ist, hat $x_4 = t$ Perioden zurückgelegt beim Eintreten des Punktereignisses.[*]

Diese Auffassung von Raum und Zeit schwebte den Physikern stets, wenn auch meist unbewusst, vor, wie aus der Rolle klar erkennbar ist, welche diese Begriffe in der messenden Physik spielen; diese Auffassung musste der Leser auch der zweiten Betrachtung des letzten § zugrunde legen, um mit diesen Ausführungen einen Sinn verbinden zu können. Aber wir wollen nun zeigen, dass man sie fallen lassen und durch eine allgemeinere ersetzen muss, um das Postulat der allgemeinen Relativität durchführen zu können, falls die spezielle Relativitätstheorie für den Fall des Fehlens eines Gravitationsfeldes zutrifft.

Wir führen in einem Raume, der frei sei von Gravitationsfeldern, ein Galileisches Bezugssystem $K(x, y, z, t)$ ein, und außerdem ein relativ zu $K$ gleichförmig rotierendes Koordinatensystem $K'(x', y', z', t')$. Die Anfangspunkte beider Systeme sowie deren $Z$-Achsen mögen dauernd zusammenfallen. Wir wollen zeigen, dass für eine Raum-Zeitmessung im System $K'$ die obige Festsetzung für die physikalische Bedeutung von Längen und Zeiten nicht aufrecht erhalten werden kann. Aus Symmetriegründen ist klar, dass ein Kreis um den Anfangspunkt in der $X$-$Y$-Ebene von $K$ zugleich ein Kreis in der $X'$-$Y'$-Ebene von $K'$ ist. Wir denken uns nun Umfang und Durchmesser dieses Kreises mit einem (relativ zum Radius unendlich kleinen) Einheitsmaßstab ausgemessen und den Quotienten beider Messresultate gebildet. Würde man dies Experiment mit einem relativ zum Galileischen System $K$ ruhenden Maßstabe ausführen, so würde man als Quotienten die Zahl $\pi$ erhalten. Das Resultat der mit einem relativ zu $K'$ ruhenden Maßstabe ausgeführten Bestimmung würde eine Zahl sein, die grösser ist als $\pi$. Man erkennt dies leicht, wenn man den ganzen Messprozess vom „ruhenden" System $K$ aus beurteilt und berücksichtigt, dass der peripherisch angelegte Maßstab eine Lorentz-Verkürzung erleidet, der radial angelegte Maßstab

---

[*] Die Konstatierbarkeit der „Gleichzeitigkeit" für räumlich unmittelbar benachbarte Ereignisse, oder – präziser gesagt – für das raumzeitliche unmittelbare Benachbart-Sein (Koinzidenz) nehmen wir an, ohne für diesen fundamentalen Begriff eine Definition zu geben.

### Why does Einstein's theory of gravitation require non-Euclidean geometry?

From a historical perspective the more interesting question is, when and how did Einstein first realize that his new theory involved non-Euclidean geometry? Although he had found the equivalence principle in 1907, it was only in 1911 that he worked out its consequences and began to develop a new theory of gravitation.

In 1912 Einstein first elaborated a theory of the static gravitational field, analogous to the familiar Newtonian theory and to the case of a static electrical field. But he knew that his theory would also involve dynamical effects of gravitation, analogous to dynamical electromagnetic fields such as induction or waves; however, such dynamical effects were not known for gravitation. It is here that the equivalence principle again saved him because it suggested considering the forces of acceleration occurring in a rotating reference frame (like in Newton's rotating bucket) as the analogue to the case of a magnetic field in electrodynamics. But the case of rotation also opened new vistas to the foundations of the new theory. For Einstein the new theory of gravitation should allow a rotating frame of reference to be interpreted as being at rest and the forces occurring in it as a dynamical gravitational field.

To demonstrate why "a more general view" of space and time measurements is required, as claimed on page 2, Einstein discusses the case of a system of reference K′ that rotates with uniform rotational velocity relative to a Galilean system K, like a merry-go-around. He compares measurements of the circumference and diameter of a rotating circular disk, using measuring rods at rest in system K, with such measurements using rods at rest in system K′.

For an observer at rest relative to the Galilean system of reference K, the ratio of the circumference to the diameter of the rotating disk, measured with rods at rest with respect to K, is $\pi$. If that observer performs the same measurements with measuring rods at rest relative to K′, the result is different. The rods used to measure the circumference suffer, according to special relativity, a Lorentz contraction with respect to that observer; therefore, more rods are needed. The length of the rods used to measure the diameter is unaffected; therefore, the ratio of the circumference to the diameter is greater than $\pi$. Thus, Euclidean geometry does not apply to the rotating system. Because of the equivalence between rotational acceleration and gravitation, Euclidean geometry has to be abandoned in a relativistic theory of gravitation.

Two of the characteristic features of Euclidean geometry are (1) the sum of angles in a triangle is 180° (or $\pi$ radians), and (2) the ratio of the circumference of a circle to its diameter is $\pi$. This is no longer necessarily the case in a non-Euclidean geometry. The distinction between the different geometries is easy to visualize in two dimensions, considering, for example, a plane and a curved surface such as a sphere. The study of non-Euclidean geometry in curved spaces was pioneered by the mathematician Gauss in the nineteenth century, and Einstein learned about it while he was a student.

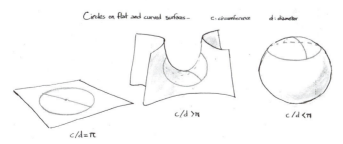

Circles on flat and curved surfaces — c: circumference   d: diameter

$c/d = \pi$    $c/d > \pi$    $c/d < \pi$

aber nicht. Es gilt daher inbezug auf $K'$ nicht die Euklidische Geometrie; der oben festgelegte Koordinatenbegriff, welcher die Gültigkeit der Euklidischen Geometrie voraussetzt, versagt also mit Bezug auf das System $K'$. Ebensowenig kann man in $K'$ eine Zeit einführen, welche durch relativ zu $K'$ ruhende, gleich beschaffene Uhren angezeigt wird. Um dies einzusehen, denke man sich im Koordinatenursprung und an der Peripherie des Kreises je eine Uhr angeordnet und vom "ruhenden" System $K$ aus betrachtet. Nach einem bekannten Resultat der speziellen Relativitätstheorie geht – von $K$ aus beurteilt – die auf der Kreisperipherie angeordnete Uhr langsamer als die im Anfangspunkt angeordnete Uhr, weil erstere Uhr bewegt ist, die letztere aber nicht. Ein im Koordinatenursprung befindlicher Beobachter, welcher auch die an der Peripherie befindliche Uhr zu beobachten fähig wäre, würde also die an der Peripherie angeordnete Uhr langsamer gehen sehen als die neben ihm angeordnete Uhr. Da er sich nicht dazu entschliessen wird, die Lichtgeschwindigkeit auf dem in Betracht kommenden Wege explizite von der Zeit abhängen zu lassen, wird er nicht anders die Zeit inbezug auf seine Beobachtung interpretieren, als dass die Uhr an der Peripherie "wirklich" langsamer gehe als die im Ursprung angeordnete. Er wird also nicht umhin können, die Zeit so zu definieren, dass die Gang-Geschwindigkeit einer Uhr vom Orte abhängt.

Wir gelangen also zu dem Ergebnis: In der allgemeinen Relativitätstheorie können Raum und Zeit-Grössen nicht so definiert werden, dass räumliche Koordinatendifferenzen unmittelbar mit dem Einheitsmassstab, zeitliche mit einer Normaluhr gemessen werden könnten.

Das bisherige Mittel, in das zeiträumliche Kontinuum in bestimmter Weise Koordinaten zu legen, versagt also, und es scheint sich kein anderer Weg darzubieten, der gestatten würde, der vierdimensionalen Welt Koordinatensysteme so anzupassen, dass bei ihrer Verwendung eine besonders einfache Gestalt der Naturgesetze zu erwarten wäre. Es bleibt daher nichts übrig, als alle denkbaren $^{\times}$ Koordinatensysteme als für die Naturbeschreibung prinzipiell gleichberechtigt anzusehen. Dies kommt auf die Forderung hinaus:

Die allgemeinen Naturgesetze sind durch Gleichungen auszudrücken, die für alle Koordinatensysteme gelten, d. h. die beliebigen Substitutionen gegenüber kovariant (allgemein kovariant) sind.

Es ist klar, dass eine Physik, welche diesem Postulat genügt, dem allgemeinen Relativitäts-Postulat gerecht wird. Denn in allen Substitutionen sind jedenfalls auch diejenigen enthalten, welche allen Relativbewegungen der (dreidimensionalen) Koordinatensysteme entsprechen.

$^{\times}$ Von gewissen Beschränkungen, welche der Forderung der eindeutigen Zuordnung und derjenigen der Stetigkeit entsprechen, wollen wir hier nicht sprechen.

## What is the role of coordinates in the new theory of gravitation?

To define the position of a point in space, one has to choose a system of coordinates. A convenient choice of coordinates in a plane is the Cartesian coordinate system, a grid of straight and perpendicular lines. Alternatively, one could choose a curvilinear system of coordinates, but in a flat space this can always be transformed to Cartesian coordinates. On a curved surface not every choice of a coordinate system can be reduced to the Cartesian form.

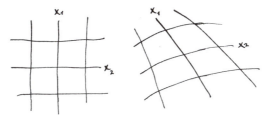

The discussion on this and the following pages is the prelude to what would emerge as the central idea of general relativity, namely, that gravity can be considered as the geometry of curved spacetime. In January 1921, Einstein gave a lecture before the Prussian Academy of Sciences in which he described the role of geometry in physics. He asserted there: "Geometry thus completed is evidently a natural science; we may in fact regard it as the most ancient branch of physics. Its affirmations rest essentially on induction from experience, but not on logical inferences only. We call this 'practical geometry.' . . . The question whether the practical geometry of the universe is Euclidean or not has a clear meaning, and its answer can only be furnished by experience."

The question of the physical significance of coordinates remained a puzzle for Einstein until the very completion of his new theory of gravitation. In classical physics, their significance is intuitively clear: they serve to identify and label events in space and time with the help of rulers and clocks. But the consideration of a rotating frame of reference and its interpretation in terms of a dynamical gravitational field showed that this simple relation between coordinates and spacetime measurements cannot hold.

After having considered measurement rods, Einstein discusses the strange behavior of clocks in a gravitational field. For this purpose, he again uses the mental model of a frame of reference K′ rotating with respect to a frame at rest K. An observer at the common origin of coordinates compares (by means of a light beam) the time on her clock with the time shown by a clock at the circumference of the rotating disk. Due to the time dilation of special relativity, she finds that the clock at the circumference is lagging her clock. The amount of delay depends on the velocity of the clock on the rotating disk, namely, on its distance from the origin. The time measured by a clock at rest with respect to a rotating frame of reference depends on where the clock may be. According to the equivalence principle, it follows that the rate of a clock in a gravitational field must depend on the position of the clock. Einstein concludes that, just as spatial coordinate differences cannot directly be measured with a unit rod, temporal coordinate differences cannot directly be measured by a normal clock.

The failure to ascribe a direct physical meaning to a system of coordinates led Einstein to the conclusion that all imaginable systems of coordinates are, in principle, suitable for the description of the laws of nature. Therefore, the equations that represent these laws must remain intact with respect to any transformation of coordinates. They must be *generally covariant*. Einstein uses here, for the first time in this manuscript, the concept of general covariance.

(2)

Dass diese Forderung der allgemeinen Kovarianz, welche dem Raume und der Zeit den letzten Rest physikalischer Gegenständlichkeit nehmen, eine natürliche Forderung ist, geht aus folgender Überlegung hervor. Alle unsere zeiträumlichen Konstatierungen laufen stets auf die Bestimmung zeiträumlicher Koinzidenzen hinaus. Bestände beispielsweise das Geschehen nur in der Bewegung materieller Punkte, so wäre letzten Endes nichts beobachtbar als die Begegnungen zweier oder mehrerer dieser Punkte. Auch die Ergebnisse unserer Messungen sind nichts anderes als die Konstatierung derartiger Begegnungen materieller Punkte unserer Massstäbe mit andern materiellen Punkten, bezw. Koinzidenzen zwischen Uhrzeigern, Zifferblattpunkten und ins Auge gefassten am gleichen Orte und zur gleichen Zeit stattfindenden Punktereignissen.

Die Einführung eines Bezugssystems dient zu nichts anderem als zur leichteren Beschreibung der Gesamtheit solcher Koinzidenzen. Man ordnet der Welt vier zeiträumliche Variable $x_1, x_2, x_3, x_4$ zu derart, dass jedem Punktereignis ein Wertesystem der Variabeln $(x_1 .. x_4)$ entspricht. Zwei koinzidierenden Punktereignissen entspricht dasselbe Wertsystem der Variabeln $x_1 .. x_4$; d. h. die Koinzidenz ist durch die Übereinstimmung der Koordinaten charakterisiert. Führt man statt der Variabeln $x_1 .. x_4$ beliebige Funktionen derselben, $x_1', x_2', x_3', x_4'$ als neues Koordinatensystem ein, sodass die Wertesysteme einander eindeutig zugeordnet sind, so ist die Gleichheit aller vier Koordinaten auch im neuen System der Ausdruck für die raum-zeitliche Koinzidenz zweier Punktereignisse. Da sich alle unsere physikalischen Erfahrungen letzten Endes auf solche Koinzidenzen zurückführen lassen, ist kein Grund vorhanden gewisse Koordinatensysteme vor anderen zu bevorzugen, d. h. wir gelangen zu der Forderung der allgemeinen Kovarianz.

§ 4. Die fundamentale Masseigenschaft

Beziehung der vier Koordinaten zu räumlichen und zeitlichen Messergebnissen. Analytischer Ausdruck für das Gravitationsfeld.

Es kommt mir in dieser Abhandlung nicht darauf an, die allgemeine Relativitätstheorie als ein möglichst einfaches logisches System mit einem Minimum von Axiomen darzustellen. Sondern es ist mein Hauptziel, diese Theorie so zu entwickeln, dass der Leser die psychologische Natürlichkeit des eingeschlagenen Weges empfindet, und dass die zu Grunde gelegten Voraussetzungen durch die Erfahrung möglichst gesichert erscheinen. In diesem Sinne sei nun die Voraussetzung eingeführt:

Für unendlich kleine vierdimensionale Gebiete ist die Relativitätstheorie im engeren Sinne bei passender Koordinatenwahl zutreffend. (unendlich kleinen ("örtlichen"))

Der Beschleunigungs-Zustand des Koordinatensystems ist hierbei so zu wählen, dass ein Gravitationsfeld nicht auftritt; dies ist für ein unendlich kleines Gebiet möglich. $X_1, X_2, X_3$ seien die räumlichen Koordinaten; $X_4$ die zugehörige

### What is the meaning of general covariance?

For Einstein, the general covariance of his theory had a deep philosophical meaning. It was the mathematical expression of the generalized principle of relativity and thus the cornerstone of his theory. This idea was later challenged on the grounds that general covariance is only a mathematical property of the equations and not a physical property of nature. It turned out that any theory can be formulated in a generally covariant way. Indeed, Newtonian mechanics and special relativity also can be given generally covariant formulations. However, these formulations do not get rid of a fixed spacetime background in the way general relativity does.

In elaborating his theory, Einstein at first believed that coordinates could be used by themselves to identify events in space and time and thus have a physical meaning. But this idea led to trouble, as it seemed to suggest that his theory could not be generally covariant. In fact, by assuming that coordinates carry a physical meaning by themselves, he was able in 1913 to construct two distinct solutions to the same field equations using an argument that came to be known as the "hole argument" because it refers to a matter-free region of spacetime in which the two distinct solutions supposedly coexist. From a given solution, a new solution is constructed within the hole by distorting the original coordinate system and ascribing the value of the field at some point in the new coordinate system to a point that has the same coordinate labels in the old system (see the figure). At the time, this argument helped Einstein justify the fact that the preliminary *Entwurf* theory he had formulated with Grossmann was indeed not generally covariant.

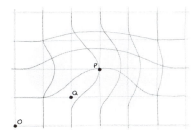

But after Einstein formulated the theory of general relativity in 1915, which does not display such problems, he still had to get rid of this hole argument. In the course of examining this point, Einstein realized that the coordinates do not have an intrinsic physical meaning. It was probably the philosopher Moritz Schlick who suggested to him that only physically measurable spacetime coincidences, such as the crossing of two light rays or the position of a clock hand, have physical meaning.

This is the complex train of thought that led Einstein to write at the beginning of this page: "That this requirement of general co-variance, which takes away from space and time the last remainder of physical objectivity, is a natural one, will be seen from the following reflection. All our space-time verifications invariably amount to a determination of space-time coincidences."

Note that Einstein changed the title of section 4 from "The Fundamental Metric Property" to the more descriptive title "The Relation of the Four Co-ordinates to Measurement in Space and Time," which corresponds to his emphasis in the first sentence of the following page that the main objective was not to give a logically succinct presentation of the theory but to make it intuitively plausible.

in geeignetem Massstab gemessene Zeitkoordinate. Diese Koordinaten haben, wenn ein starres Stäbchen als Einheitsmassstab gegeben gedacht wird, beigegebner Orientierung des Koordinatensystems unmittelbare physikalische Bedeutung im Sinne der speziellen Relativitätstheorie. Der Ausdruck

$$ds^2 = -dX_1^2 - dX_2^2 - dX_3^2 + dX_4^2 \quad \cdots (1)$$

hat dann nach der speziellen Relativitätstheorie einen von der Orientierung des lokalen Koordinatensystems unabhängigen, durch Raum-Zeitmessung ermittelbaren Wert. Wir nennen ds die Grösse des zu den unendlich benachbarten Punkten der vierdimensionalen Raumes gehörigen Zeitelementes. Ist das zu dem Element $(dX_1 \ldots dX_4)$ gehörige $ds^2$ positiv, so nennen wir mit Minkowski ersteres zeitartig, im entgegengesetzten Falle raumartig.

Zu dem betrachteten Linienelemente gehören auch bestimmte Differentiale $dx_1 \ldots dx_4$ der vierdimensionalen Koordinaten des gewählten Bezugssystems. Ist dieses, sowie ein „lokales" System obiger Art gegeben, so werden sich hier die $dX_\nu$ durch bestimmte lineare homogene Ausdrücke der $dx_\sigma$ darstellen lassen:

$$dX_\nu = \sum_\sigma \alpha_{\nu\sigma} \, dx_\sigma \quad \cdots \cdots (2)$$

Setzt man diese Ausdrücke in (1) ein, so erhält man

$$ds^2 = \sum_{\sigma\tau} g_{\sigma\tau} \, dx_\sigma \, dx_\tau, \quad \cdots \cdots (3)$$

wobei die $g_{\sigma\tau}$ Funktionen der $x_\sigma$ sein werden, die nicht mehr von der Orientierung und dem Bewegungszustand des „lokalen" Koordinatensystems abhängen können, denn $ds^2$ ist eine durch Massstab- Uhren- Messung ermittelbare, zu den betrachteten, zeiträumlich unendlich benachbarten Punktereignissen gehörige, unabhängig von jeder besonderen Koordinatenwahl definierte Grösse. Die $g_{\sigma\tau}$ sind hierbei so zu wählen, dass $g_{\sigma\tau} = g_{\tau\sigma}$ ist; die Summation ist über alle Werte von $\sigma$ und $\tau$ zu erstrecken, sodass die Summe aus $4 \times 4$ Summanden besteht, von denen 12 paarweise gleich sind.

Der Fall der gewöhnlichen Relativitätstheorie geht aus dem hier betrachteten Spezialisierung hervor, falls es möglich ist, vermöge des besonderen Verhaltens der $g_{\sigma\tau}$ in einem endlichen Gebiete, in diesem das Bezugssystem so zu wählen, dass die $g_{\sigma\tau}$ die konstanten Werte

$$\left.\begin{array}{cccc} -1 & 0 & 0 & 0 \\ 0 & -1 & 0 & 0 \\ 0 & 0 & -1 & 0 \\ 0 & 0 & 0 & +1 \end{array}\right\} (4)$$

x Im „Lichtvakuum" gemessene Zeit. Die Zeiteinheit ist so zu wählen, dass die Vakuum- Geschwindigkeit des Lichtes gleich 1 wird. — in dem „lokalen" Koordinatensystem gemessen— gleich 1 wird.

## What is the geometry of spacetime?

In 1907, Minkowski developed a mathematical formalism for representing physical events in space and time and the relations between these events as implied by the special theory of relativity. The mathematical formalism consists of a four-dimensional spacetime, in which coordinate systems representing physical frames of reference are used to characterize each physical event by four numbers: its three space coordinates and one time coordinate. The time coordinate is always multiplied by the constant speed of light, and thus it also acquires the dimension of a spatial coordinate, so that Minkowski's four-dimensional world becomes even more similar to the three-dimensional Euclidean world of classical physics.

On this page, Einstein introduces the concepts of the "line element" and the "metric tensor." The line element $ds$ ($d$ stands for an infinitesimally small difference) is the distance between two adjacent points in spacetime. It is expressed in terms of the projections of the segment connecting the two points on the lines defining the grid of coordinates—$dx_\mu$ ($\mu = 1,2,3,4$). In the flat Minkowski spacetime of special relativity, the square of the line element is given by eq. (1). It is essentially the extension of the Pythagorean theorem to four dimensions and adaptation to the special nature of the time coordinate.

In curved four-dimensional spacetime, 10 numbers are needed to calculate the distance from one point to any neighboring point, and the expression for $ds^2$ is now given by eq. (3). It is convenient to present these numbers in a 4x4 matrix array $g_{\mu\nu}$, where the first index represents the line, and the second, the column of this matrix. This array is the "metric tensor," which reflects the geometrical properties of spacetime in the chosen coordinate system. Its components, in general, are functions of position in spacetime. The metric tensor has 16 components, but only 10 of them are independent because of the symmetry between the off-diagonal elements,—$g_{12} = g_{21}, \dots$. In the flat spacetime of special relativity, the metric tensor reduces to the array of eq. (4). In accelerated frames of reference and, hence, in the presence of gravitational fields, there will always be nonconstant elements of the metric tensor.

$$\begin{pmatrix} g_{11} & g_{12} & g_{13} & g_{14} \\ g_{21} & g_{22} & g_{23} & g_{24} \\ g_{31} & g_{32} & g_{33} & g_{34} \\ g_{41} & g_{42} & g_{43} & g_{44} \end{pmatrix}$$

After Einstein had unsuccessfully attempted to incorporate gravitation into his special theory of relativity, he decided to base a relativistic theory of gravitation on the equivalence principle, stating that a static gravitational field can be simulated by a reference frame in uniform acceleration. This consideration allowed him to use the special relativistic effects in such a moving reference frame to draw conclusions about gravitation. He first published the principle in a review article on special relativity in 1907. There he also discussed some of its immediate implications, such as the bending of light rays in a gravitational field and the effect of gravitation on the pace of clocks. It was not until four years later, when Einstein was serving as professor of theoretical physics in the German section of the Charles University in Prague (April 1911–July 1912), that he gave a more complete formulation of the principle and elaborated its consequences in greater detail.

(für endliche Gebiete)

annehmen. Wir werden später sehen, dass die Wahl solcher Koordinaten im Allgemeinen nicht möglich ist.

Aus den Betrachtungen des §2 und §3 geht hervor, dass die Grössen $g_{\sigma\tau}$ vom physikalischen Standpunkte aus als diejenigen Grössen anzusehen sind, welche das Gravitationsfeld inbezug auf das gewählte Bezugssystem beschreiben. Nehmen wir nämlich zunächst an, es sei für ein betrachtetes gewisses (vierdimensionales) Gebiet bei geeigneter Wahl der Koordinaten die spezielle Relativitätstheorie gültig. Die $g_{\sigma\tau}$ haben dann die in (4) angegebenen Werte. Ein freier materieller Punkt bewegt sich dann bezüglich dieses Systems geradlinig gleichförmig. Führt man nun durch eine beliebige Substitution neue Raum-Zeit-Koordinaten $x_1 \ldots x_4$ ein, so werden in diesem neuen System die $g_{\sigma\tau}$ nicht mehr Konstante sondern Raum-Zeit-Funktionen sein. Gleichzeitig wird sich die Bewegung des freien Massenpunktes in den neuen Koordinaten als eine krummlinige, nicht gleichförmige, darstellen, wobei dies Bewegungsgesetz unabhängig sein wird von der Natur des bewegten Massenpunktes. Wir werden also diese Bewegung als eine solche unter dem Einfluss eines Gravitationsfeldes deuten. Wir sehen also das Auftreten eines Gravitationsfeldes geknüpft an eine raumzeitliche Veränderlichkeit der $g_{\sigma\tau}$. Auch in dem Falle, dass wir nicht in einem endlichen Gebiete bei passender Koordinatenwahl die Gültigkeit der speziellen Relativitätstheorie herbeiführen können, werden wir an der Auffassung festhalten, dass die $g_{\sigma\tau}$ das Gravitationsfeld beschreiben.

Die Gravitation spielt also gemäss der allgemeinen Relativitätstheorie eine Ausnahmerolle gegenüber den übrigen, insbesondere den elektromagnetischen Kräften, indem die das Gravitationsfeld darstellenden 10 Funktionen $g_{\sigma\tau}$ zugleich die metrischen Eigenschaften des vierdimensionalen Messraumes bestimmen.

B. Mathematische Hülfsmittel für die Aufstellung allgemein kovarianter Gleichungen.

Nachdem wir im übrigen gesehen haben, dass das allgemeine Relativitäts-Postulat zu der Forderung führt, dass die Gleichungssysteme der Physik beliebigen Substitutionen der Koordinaten $x_1 \ldots x_4$ gegenüber kovariant sein müssen, haben wir zu überlegen, wie derartige allgemein kovariante Gleichungen gewonnen werden können. Dieser rein mathematischen Aufgabe wenden wir uns jetzt zu. Es wird sich dabei zeigen, dass bei der Lösung die in Gleichung (3) angegebene Invariante $ds$ eine fundamentale Rolle spielt, welche wir in Anlehnung an die Gauss'sche Flächentheorie als „Linienelement" bezeichnet haben.

§5. Die geodätische Linie (Gleichungssystem der Punktbewegung).
Der Grundgedanke dieser allgemeinen Kovariantentheorie ist folgender. Es seien gewisse Dinge („Tensoren") inbezug auf jedes Koordinaten-

### When did Einstein realize that gravitation has to be described by a complex mathematical expression?

In 1911, in Prague, Einstein focused on developing a consistent theory of the static gravitational field based on the equivalence principle. Like Newton's theory of gravity, Einstein's theory involved a gravitational potential represented by a single scalar function, now given by a variable speed of light. Some basic features of the final theory of general relativity had already been conceived by then, among which was the understanding that the source of the gravitational potential is not only the mass of concrete bodies but also the equivalent mass of the energy of the gravitational field itself. Thus, the gravitational field, generated by the source, can act as its own source, and the field equation is bound to be nonlinear. However, as long as Einstein restricted himself to considering a static gravitational field, he continued to assume that the gravitational potential was represented by a single function.

When Einstein tried to generalize his preliminary theory of the static gravitational field in 1912 during his Prague period, he realized that gravitation must be described by a much more complicated mathematical object than in classical physics. Instead of being a single scalar function, the gravitational potential in his theory is described by a "metric tensor," a mathematical object with 10 independent functions. At the same time, this object describes the geometry of a four-dimensional spacetime. On this basis, gravitation can be conceived of as a geometric property of spacetime. It would take a long time, however, to spell out all the consequences of this implication.

The discovery that the gravitational potential is represented by the metric of spacetime is one of the most important landmarks on the road to general relativity. One finds a hint at how this break-through occurred in the last sentence, in a note added in proof, to the last paper that Einstein wrote on gravitation in Prague: "The Hamiltonian equation, which was the last one written down, gives an idea about how the equations of motion of a material point in a dynamic gravitational field are constructed." In fact, Einstein had managed to write an equation of his theory of the static field in such a way that it suggested a generalization involving the metric tensor.

In part B, Einstein will do what he promised in the introduction, namely, to develop all the necessary tools so that a study of mathematics would not be required to understand this paper. Initially, he meant to begin his exposition by describing the basic elements of the geometry of curved space: the geodetic line. Then, he decided that some mathematical preparation is required and crossed out the line with the intended title of section 5, deferring it to section 10.

system definiert durch eine Anzahl Raumfunktionen, welche die "Komponenten" des Tensors genannt werden. Es gibt dann gewisse Regeln, nach welchen diese Komponenten für ein neues Koordinatensystem berechnet werden, wenn sie für das ursprüngliche System bekannt sind, und wenn die beide Systeme verknüpfende Transformation bekannt ist. Die späterhin als Tensoren bezeichneten Dinge sind ferner dadurch gekennzeichnet, dass die Transformationsgleichungen für ihre Komponenten linear und homogen sind. Demnach verschwinden sämtliche Komponenten im neuen System, wenn sie im ursprünglichen System sämtliche verschwinden. Wird also ein Naturgesetz durch das Nullsetzen aller Komponenten eines Tensors formuliert, so ist es allgemein kovariant. Indem wir also die Bildungsgesetze der Tensoren untersuchen, erlangen wir die Mittel zur Aufstellung allgemein kovarianter Gesetze.

### §5. Kontravarianter und kovarianter Vierervektor.

**Kontravarianter Vierervektor.** Das Linienelement ist definiert durch die vier "Komponenten" $dx_\nu$, deren Transformationsgesetz durch die Gleichung

$$dx_\sigma' = \sum_\nu \frac{\partial x_\sigma'}{\partial x_\nu} dx_\nu \quad \ldots \ldots (5)$$

Die $dx_\sigma'$ drücken sich linear und homogen durch die $dx_\nu$ aus, wir können diese Koordinatendifferentiale $dx_\nu$ daher als die Komponenten eines "Tensors" ansehen, den wir speziell als kontravarianten Vierervektor bezeichnen. Jedes Ding, was bezüglich des Koordinatensystems durch vier Grössen $A^\nu$ definiert ist, die sich nach dem Gesetz

$$A^\sigma' = \sum_\nu \frac{\partial x_\sigma'}{\partial x_\nu} A^\nu \quad \ldots \ldots (5a)$$

transformieren bezeichnen wir als kontravarianten Vierervektor. Aus (5a) folgt sogleich, dass die Summen $(A^\sigma \pm B^\sigma)$ ebenfalls Komponenten eines Vierervektors sind, wenn $A^\sigma$ und $B^\sigma$ es sind. Entsprechendes gilt für alle später als "Tensoren" einzuführenden Systeme (Regel von der Addition und Subtraktion der Tensoren).

**Kovarianter Vierervektor.** Vier Grössen $A_\nu$ nennen wir die Komponenten eines kovarianten Vierervektors, wenn für jede beliebige Wahl des kontravarianten Vierervektors $B^\nu$

$$\sum_\nu A_\nu B^\nu = \text{Invariante} \ldots \ldots (6)$$

Aus dieser Definition folgt das Transformationsgesetz des kovarianten Vierervektors. Ersetzt man nämlich auf der rechten Seite der Gleichung

$$\sum_\sigma A_\sigma' B^\sigma' = \sum_\nu A_\nu B^\nu$$

$B^\sigma$ gemäss (5a) durch $\sum_\nu \frac{\partial x_\sigma'}{\partial x_\nu} B^\nu$, so erhält man

$$\sum_\nu B_\nu \quad$$

## Why tensors, vectors, scalars?

Tensors are rather complicated mathematical objects. Why are they indispensable in general relativity? The laws of physics are formulated as mathematical equations in which the physical entities represented on both sides of these equations are functions of position in space and time and assume different values for observers in different reference frames, which are represented by different coordinate systems. The requirement of general covariance means that even though the two sides of an equation may change on transition from one reference frame to another, their equality is maintained for all observers, irrespective of their relative motion. The mathematical entities with this property are tensors. Vectors and scalars are special cases of tensors.

> While vectors and vector analysis were known to physicists through the triumph of electromagnetism, tensors were known to only a few specialists working in fields like crystallography. It was no surprise, then, that when Einstein realized he would require more sophisticated mathematical methods than he was familiar with to make progress, he turned to his friend the mathematician: "Grossman, you have got to help me or I will go crazy." Together they plunged into the absolute differential calculus of Riemann, Ricci, and Levi-Civita, leading from Gaussian geometry of surfaces in a three-dimensional space to Riemannian geometry in higher dimensions.
>
> Einstein then wrote to the physicist Arnold Sommerfeld: "I am now working exclusively on the gravitation problem and I believe that, with the help of a mathematical friend here, I will overcome all difficulties. . . . I have gained enormous respect for mathematics, whose more subtle parts I considered until now, in my ignorance, as luxury! Compared to this problem, the original theory of relativity is a child's game."

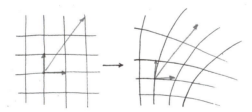

Einstein explains the motivation for presenting a comprehensive account of tensor calculus and begins with the basic definitions. The simplest tensors are vectors (tensors of rank 1) and scalars (tensors of rank 0). Vectors are objects that have a magnitude and direction at every point in space (later, together with Einstein, we shall confine our discussion to spacetime) and are represented by a set of components related to the basis vectors defined by the coordinate system chosen to describe the points in space. Vectors are denoted by letters of the Greek alphabet, and their number is the dimension of space, thus in our case $\mu = 1,2,3,4$. In shifting from one coordinate system, $x_\mu$, to another, $x'_\mu$, the components of a vector change— because the basis vectors change—as defined by a transformation rule. Contravariant vectors (conventionally denoted by upper indices) differ from covariant vectors (denoted by lower indices) by how their components are determined by the basis vectors and are characterized by different transformation rules. The sum of products of the components of a contravariant and a covariant vector (eq. 6) is a function of position in spacetime, which is invariant under coordinate transformations. Such functions are called scalars or tensors of rank 0.

## 11

$B_{\chi}^{\nu}$ durch den aus der Umkehrung der Gleichung (5a) folgenden Ausdruck $\sum_{\sigma} \frac{\partial x_\nu}{\partial x_\sigma'} B^{\sigma'}$, so erhält man

$$\sum_{\sigma} B^{\sigma'} \sum_{\nu} \frac{\partial x_\nu}{\partial x_\sigma'} A_\nu = \sum_{\sigma} B^{\sigma'} A_\sigma'.$$

Hieraus folgt aber, weil in dieser Gleichung die $B^{\sigma'}$ unabhängig voneinander frei wählbar sind, das Transformationsgesetz

$$A_\sigma' = \sum_{\nu} \frac{\partial x_\nu}{\partial x_\sigma'} A_\nu \quad \cdots\cdots (7).$$

Bemerkung zur Vereinfachung der Schreibweise der Ausdrücke. Ein Blick auf die Gleichungen dieses § zeigt, dass über Indizes, die zweimal unter einem Summenzeichen auftreten (z. B. der Index $\nu$ in (5)) stets summiert wird, und zwar nur über zweimal auftretende Indizes. Es ist deshalb möglich, ohne die Klarheit zu beeinträchtigen, die Summenzeichen wegzulassen. Dafür führen wir die Vorschrift ein: Tritt ein Index in einem Term eines Ausdruckes zweimal auf, so ist über ihn stets zu summieren, wenn nicht ausdrücklich das Gegenteil bemerkt ist.

Der Unterschied zwischen dem kovarianten und kontravarianten Vierervektor liegt in dem Transformationsgesetz ((7) bezw. (5)). Beide Gebilde sind Tensoren im Sinne der obigen allgemeinen Bemerkung; hierin liegt ihre Bedeutung. Im Anschluss an Ricci und Levi-Cività wird der kontravariante Charakter durch oberen, der kovariante durch unteren Index bezeichnet.

§ 6. Tensoren zweiten und höheren Ranges.

Kontravarianter Tensor. Bilden wir sämtliche 16 Produkte $A^{\mu\nu}$ der Komponenten $A^\mu$ und $B^\nu$ zweier kontravarianten Vierervektoren

$$A^{\mu\nu} = A^\mu B^\nu \quad \cdots\cdots (8),$$

so erfüllt $A^{\mu\nu}$ gemäss (8) und (5a) das Transformationsgesetz

$$A^{\sigma\tau'} = \frac{\partial x_\sigma'}{\partial x_\mu} \frac{\partial x_\tau'}{\partial x_\nu} A^{\mu\nu} \quad \cdots\cdots (9).$$

Wir nennen ein Ding, das bezüglich eines Bezugssystems durch 16 Grössen (Funktionen) beschrieben wird, die das Transformationsgesetz (9) erfüllen, einen kontravarianten Tensor zweiten Ranges. Nicht jeder Tensor lässt sich gemäss (8) aus zwei Vierervektoren bilden. Aber es ist leicht zu beweisen, dass sich 16 beliebig gegebene $A^{\mu\nu}$ darstellen lassen als die Summe der $A^\mu B^\nu$ von vier geeignet gewählten Paaren von Vierervektoren. Deshalb kann man beinahe alle Sätze, die für den durch (9) definierten Tensor zweiten Ranges gelten, am einfachsten dadurch beweisen, dass man sie für spezielle Tensoren vom Typus (8) darthut.

Kontravarianter Tensor beliebigen Ranges. Es ist klar, dass man entsprechend (8) und (9) auch kontravariante Tensoren dritten und höheren Ranges definieren kann mit $4^3$ etc. Komponenten. Ebenso erhellt aus (8) und (9), dass man in diesem Sinne den kontravarianten Vierervektor als kontravarianten Tensor ersten Ranges auffassen kann.

Kovarianter Tensor. Bildet man andererseits die 16 Produkte $A_{\mu\nu}$ der Komponenten zweier kovarianter Vierervektoren $A_\mu$ und $B_\nu$

$$A_{\mu\nu} = A_\mu B_\nu \quad \cdots\cdots (10),$$

### When did Einstein realize that he needed more sophisticated mathematical methods?

In his introduction to the Czech edition of his "About the Special and General Theory of Relativity in Plain Terms," from which we have already quoted (p. 9), Einstein recalls: "[T]he decisive idea of the analogy between the mathematical formulation of the theory and the Gaussian theory of surfaces came to me only in 1912 after my return to Zurich, without being aware at that time of the work of Riemann, Ricci and Levi-Civita. This was first brought to my attention by my friend Grossmann. . . ." Einstein truly considered Grossmann his friend. In 1905, he dedicated his doctoral dissertation, "A New Determination of Molecular Dimension," to "My Friend Dr. Marcel Grossmann." It was Grossmann's initiative to bring Einstein back to Zurich, to the ETH, after Grossmann became the dean of mathematics in 1911. Einstein preferred to accept Grossmann's invitation over the option to go to Leiden, the Netherlands, and eventually to become Lorentz's successor there.

This "decisive idea," mentioned in the previous paragraph, marks Einstein's transition from the static to the dynamic relativistic theory of gravitation. Despite the preceding quotation, it is most likely that he already knew before going to Zurich that the scalar theory of gravitation had to be abandoned and that a more complicated geometry of spacetime was required. At the end of 1911 he corresponded about this issue with Laue.

In March 1912, Einstein wrote to his friend Besso: "Lately I have been working like mad on the gravitation problem. Now I have gotten to the stage where I am finished with the statics. I do not know anything yet about the dynamic field, that will come only now." The transition to dynamic fields would allow Einstein to demonstrate that there are no preferred frames of reference; specifically, the same laws should apply in inertial and rotating frames of reference. On this point he remarked in the same letter to Besso: "You see that I am still far from being able to conceive of rotation as rest! Each step is devilishly difficult, and what I have derived so far is certainly the simplest of all."

Einstein went on to define tensors of rank 2 denoted by two indices. Such tensors can be formed from the 16 products of the elements of two contravariant vectors (the so-called outer multiplication). In that case, a contravariant tensor of rank 2 is obtained. In a similar way, a covariant tensor, or a mixed tensor, can be formed with one contravariant (upper) and one covariant (lower) index. Tensors are characterized by how they transform when a coordinate system $x_\mu$ is replaced by another coordinate system $x'_\mu$.

Up to eq. (7) Einstein has been using the summation symbol $\Sigma$ with an index below it, indicating the summation of the terms corresponding to the possible values of that index (in this case, four values). Einstein introduces the convention, since widely adopted, that whenever an index appears twice, once as a lower and once as an upper index, a summation over that index is implied, without the need for the summation symbol.

so gilt für diese das Transformationsgesetz

$$A_{\sigma\tau}' = \frac{\partial x_\mu}{\partial x_\sigma'} \frac{\partial x_\nu}{\partial x_\tau'} A_{\mu\nu}. \quad \ldots \ldots (11)$$

Durch dies Transformationsgesetz wird der kovariante Tensor zweiten Ranges definiert. Alle Bemerkungen, welche vorher über die kontravarianten Tensoren gemacht wurden, gelten auch für die kovarianten Tensoren.

Bemerkung. Es ist bequem, den Skalar (Invariante) sowohl als kontravarianten wie als kovarianten Tensor vom Range null zu behandeln.

Gemischter Tensor. Man kann auch einen Tensor vom Typus zweiten Ranges

$$A_\mu{}^\nu = A_\mu B^\nu \quad \ldots (12)$$

definieren, der bezüglich des Index $\mu$ kovariant, bezüglich des Index $\nu$ kontravariant ist. Sein Transformationsgesetz ist

$$A_\sigma{}^{\tau'} = \frac{\partial x_\sigma'}{\partial x_\beta} \frac{\partial x_\tau}{\partial x_\sigma'} A_\alpha{}^\beta. \quad \ldots (13)$$

Natürlich gibt es gemischte Tensoren mit beliebig vielen Indizes kovarianten und beliebig vielen Indizes kontravarianten Charakters. Der kovariante und der kontravariante Tensor können als spezielle Fälle des gemischten angesehen werden.

Symmetrische Tensoren. Ein kontravarianter bezw. kovarianter Tensor zweiten oder höheren Ranges heisst symmetrisch, wenn zwei Komponenten, die durch Vertauschung irgend zweier Indizes auseinander hervorgehen, gleiche sind. Der Tensor $A^{\mu\nu}$ bezw. $A_{\mu\nu}$ ist also symmetrisch, wenn für jede Kombination der Indizes

$$A^{\mu\nu} = A^{\nu\mu} \quad \ldots \ldots (14)$$

$$\text{bezw.} \quad A_{\mu\nu} = A_{\nu\mu} \quad \ldots \ldots (14\alpha)$$

ist.

Es muss bewiesen werden, dass die so definierte Symmetrie eine vom Bezugsystem unabhängige Eigenschaft ist. Aus (9) folgt in der That mit Rücksicht auf (11)

$$A^{\sigma\tau'} = \frac{\partial x_\sigma'}{\partial x_\mu} \frac{\partial x_\tau'}{\partial x_\nu} A^{\mu\nu} = \frac{\partial x_\sigma'}{\partial x_\mu} \frac{\partial x_\tau'}{\partial x_\nu} A^{\nu\mu} = \frac{\partial x_\tau'}{\partial x_\mu} \frac{\partial x_\sigma'}{\partial x_\nu} A^{\mu\nu} = A^{\tau\sigma'}$$

Die vorletzte Gleichsetzung beruht auf der Vertauschung der Summations-Indizes $\mu$ und $\nu$ (d. h. auf blosser Aenderung der Bezeichnungsweise).

Antisymmetrische Tensoren. Ein kontravarianter bezw. kovarianter Tensor zweiten dritten oder vierten Ranges heisst antisymmetrisch, wenn zwei Komponenten die durch Vertauschung irgend zweier Indizes auseinander hervorgehen entgegengesetzt gleich sind. Der Tensor $A^{\mu\nu}$ bezw. $A_{\mu\nu}$ ist also antisymmetrisch, wenn stets

$$A^{\mu\nu} = -A^{\nu\mu} \quad \ldots \ldots (15)$$

$$\text{bezw.} \quad A_{\mu\nu} = -A_{\nu\mu} \quad \ldots \ldots (15\alpha)$$

ist.

## What lessons could be learned from the theory of electromagnetism?

Newton's theory of gravitation was not sufficient to show the way to a dynamical theory in which gravitation propagates in space and time. Einstein realized at an early stage that he could be guided toward achieving this goal by the familiar theory of electromagnetism, as formulated by Lorentz. The essence of this formulation is that electromagnetism is a field theory, not confined to the interacting particles but extending to their surroundings. The model describes how the space-filling field is generated by the electric charges and currents. The electric field at a point in space is the force acting on a unit of electric charge at that point. Thus (electric) matter is considered to be the "source" of the field, and the field in turn determines how this matter moves. Likewise, the relativistic theory of gravitation is a field theory: a theory of the gravitational field. The mathematical representation of the physical processes according to such a model necessarily includes two parts:

- an equation of motion, describing the motion of particles in a given gravitational field; and
- a field equation, describing the gravitational field generated by its source—energy and matter.

Furthermore, Einstein conceived the unification of gravitation and inertia as analogous to the unification of the electric and magnetic fields that was so successful in the special theory of relativity. There only the electromagnetic field as a whole had a frame-independent meaning, not the separate electric or magnetic field. Einstein accepted the Lorentz model as a heuristic guideline, yet he soon found that the task of finding a field equation was the most difficult challenge he had to face in his search for a relativistic theory of gravitation.

The tensors of rank 2, which appear in this article, have a symmetry property. This is best demonstrated by writing the tensors as a 4×4 matrix. They can be either symmetric, in which case components on both sides of the diagonal are equal ($A_{12} = A_{21}, \ldots$), or antisymmetric, in which case they are of opposite sign ($A_{12} = -A_{21}, \ldots$). Symmetric tensors of rank 2 have 10 independent components. In antisymmetric tensors, the diagonal elements $A_{11}$, $A_{22}$, $A_{33}$, $A_{44}$ vanish, and 6 independent components remain. Tensors of higher rank may have such symmetry properties with respect to the interchange of any two indices.

$$\begin{pmatrix} A_{11} & A_{12} & A_{13} & A_{14} \\ A_{21} & A_{22} & A_{23} & A_{24} \\ A_{31} & A_{32} & A_{33} & A_{34} \\ A_{41} & A_{42} & A_{43} & A_{44} \end{pmatrix}$$

A symmetric tensor that plays a crucial role in Einstein's theory of gravitation combines the energy density, the density of the components of momentum, and the flow of energy and momentum into one mathematical object—the energy-momentum tensor. This tensor and its relation to the energy-momentum conservation laws will be discussed in greater detail in part C of this manuscript.

An antisymmetric tensor, which will be discussed in greater detail in part D, is the electromagnetic field tensor that combines the components of the electric field ($E_x$, $E_y$, $E_z$) and the magnetic field ($H_x$, $H_y$, $H_z$). Einstein's special theory of relativity implies that the electric and magnetic fields, separately, depend on the frame of reference. This tensor, introduced by Minkowski, has become part of the four-dimensional formulation of special relativity. It allows the Maxwell equations, which represent the physical relations between electric and magnetic fields, electric charges and electric currents, to be expressed in a simple form. The electromagnetic tensor is often written as a vector of six components and referred to as a six-vector.

## 13

Von den 16 Komponenten $A^{\mu\nu}$ verschwinden die vier Komponenten $A^{\mu}_{\mu}$, die übrigen sind paarweise entgegengesetzt gleich, sodass nur 6 numerisch verschiedene Komponenten vorhanden sind (Sechservektor). Ebenso sieht man, dass der (antisymmetrische) Tensor $A^{\mu\nu\sigma}$ (dritten Ranges) nur vier numerisch verschiedene Komponenten hat, der Tensor $A^{\mu\nu\sigma\tau}$ nur eine einzige. Symmetrische Tensoren höheren als vierten Ranges gibt es in einem Kontinuum von vier Dimensionen nicht.

### §7. Multiplikation der Tensoren.

Äussere Multiplikation der Tensoren. Man erhält aus den Komponenten eines Tensors vom Range $z$ und eines solchen vom Range $z'$ die Komponenten eines Tensors vom Range $z+z'$, indem man alle Komponenten des ersten mit allen Komponenten des zweiten paarweise multipliziert. So entstehen beispielsweise die Tensoren $T$ aus den Tensoren $A$ und $B$ verschiedener Art

$$T_{\mu\nu\sigma} = A_{\mu\nu} B_{\sigma}$$
$$T^{\alpha\beta\gamma\delta} = A^{\alpha\beta} B^{\gamma\delta}$$
$$T^{\gamma\delta}_{\alpha\beta} = A_{\alpha\beta} B^{\gamma\delta}$$

Der Beweis des Tensorcharakters der $T$ ergibt sich unmittelbar aus den Darstellungen 8), (10), (12) oder aus den Transformationsregeln (9), (11), (13). Die Gleichungen (8), (10), (12) sind selbst Beispiele äusserer Multiplikation (von Tensoren ersten Ranges).

Innere Multiplikation der Tensoren. Wir nennen den Ausdruck (6) das innere Produkt der kovarianten Vierervektors $A_{\mu}$ und des kontravarianten Vierervektors $A^{\mu}$. Analog kann durch innere Multiplikation

"Verjüngung" eines gemischten Tensors. Aus jedem gemischten Tensor kann ein Tensor von einem um zwei kleineren Range gebildet werden, indem man einen Index kovarianten und einen Index kontravarianten Charakters gleichsetzt und nach diesem Index summiert (Verjüngung). Man gewinnt so z. B. aus dem gemischten Tensor vierten Ranges $A^{\gamma\delta}_{\alpha\beta}$ den gemischten Tensor zweiten Ranges $A^{\delta}_{\beta} = A^{\alpha\delta}_{\alpha\beta} \left(= \sum_{\alpha} A^{\alpha\delta}_{\alpha\beta}\right)$ und aus diesem, abermals durch Verjüngung den Tensor nullten Ranges $A = A^{\beta}_{\beta} = A^{\alpha\beta}_{\alpha\beta}$.

$A = A^{\beta}_{\beta} = A^{\alpha\beta}_{\alpha\beta}$. Der Beweis dafür, dass das Ergebnis der Verjüngung wirklich Tensorcharakter besitzt, ergibt sich entweder aus der Tensordarstellung gemäss der Verallgemeinerung von (12) in Verbindung mit (6) oder aus der Verallgemeinerung von (13).

Innere und gemischte Multiplikation der Tensoren. Diese bestehen in der Kombination der äusseren Multiplikation mit der Verjüngung. Beispiele: Aus dem kovarianten Tensor zweiten Ranges $A_{\mu\nu}$ und dem kontravarianten Tensor ersten Ranges $B^{\sigma}$ bilden wir durch äussere Multiplikation den

## How can tensors be manipulated to produce new tensors by different tensor operations?

In these preliminary pages of introduction to tensor calculus, which mainly define concepts and provide basic proofs, Einstein is making an effort to make his presentation of the subject understandable and complete, and to flow naturally. He refers the reader to the transformation laws between different frames of reference, spread over the three previous pages, to demonstrate that products of tensors, presented in the middle of this page, are themselves tensors.

On this page, Einstein continues to educate the reader about different tensor operations. Specifically, he shows how to form tensors of higher rank by multiplying tensors of lower rank, and how to keep track of the covariant and contravariant properties in tensor multiplication. Another tensor operation is *contraction*: the summation over equal—upper and lower—indices of a given tensor. Contraction reduces the rank of a tensor by 2. The contraction of a mixed tensor of rank 2 results in a tensor of rank 0, namely, a "scalar" function of coordinates, which is invariant under coordinate transformations. Einstein then refers the reader to eq. (6) (see our remark at the end of p. 57), which is used there to show how to form an invariant function (a *scalar*) by multiplying two vectors. Here he uses this equation to define the concept of inner multiplication but decides to cross it out and, more naturally, introduces *inner multiplication* contrasted with *mixed multiplication*. Doing so, he relies on the definition of the concept of *contraction* of a mixed tensor.

## What were Einstein's heuristic guidelines in his search for a relativistic theory of gravitation?

When Einstein returned to Zurich and became familiar with the concepts and tools of Riemannian geometry and tensor calculus, he was already considering how to incorporate the following known physical principles into the mathematical framework that he was learning and constructing:

- The equivalence principle expresses the relation between gravitation and inertial forces (the fictitious forces acting on bodies in accelerated reference frames).
- The generalized relativity principle aims to eliminate such notions of classical physics as absolute space and inertial reference frame so that all reference frames are treated on equal footing.
- The conservation principle requires that the new theory conform to a generalized energy-momentum conservation law. In classical mechanics, there are three separate conservation laws of mass, energy, and momentum. In special relativity, they are combined into a single conservation law referring to the energy-momentum tensor. A theory of general relativity has to contain a generalized version of such a conservation law.
- The correspondence principle requires that the new theory can be reduced to Newton's theory under special limiting conditions, such as low velocities and weak gravitational fields.

It is worth pointing out an editorial typographic instruction for the printing process of the manuscript. Note the two vertical lines bracketing a mathematical equation in the lower part of the page. Einstein sometimes included mathematical equations within lines of text. The vertical lines are the editor's instructions to the typesetter to place such a mathematical expression on a separate line. The reader will find such vertical lines on a number of pages within this manuscript.

gemischten Tensor

$$D_{\mu\nu}^{\sigma} = A_{\mu\nu} B^{\sigma}$$

Durch Verjüngung nach den Indizes $\nu, \sigma$ entsteht der kovariante Vierervektor

$$D_\mu = D_{\mu\nu}^\nu = A_{\mu\nu} B^\nu$$

Diesen bezeichnen wir auch als inneres Produkt der Tensoren $A_{\mu\nu}$ und $B^\sigma$. Analog bildet man aus den Tensoren $A_{\mu\nu}$ und $B^{\sigma\tau}$ durch äussere Multiplikation und zweimalige Verjüngung das innere Produkt $A_{\mu\nu} B^{\mu\nu}$. Durch äussere Produktbildung und einmalige Verjüngung erhält man aus $A_{\mu\nu}$ und $B^{\sigma\tau}$ den gemischten Tensor zweiten Ranges $D_\mu^\tau = A_{\mu\nu} B^{\nu\tau}$. Man kann diese Operation passend als eine gemischte bezeichnen, denn sie ist eine äussere bezüglich der Indices $\mu$ und $\tau$, eine innere bezüglich der Indices $\nu$ und $\sigma$.

　　　Wir beweisen nun noch einen Satz, der zum Nachweis des Tensorcharakters oft verwendbar ist. Nach dem soeben Dargelegten ist $A_{\mu\nu} B^{\mu\nu}$ ein Skalar, wenn $A_{\mu\nu}$ und $B^{\sigma\tau}$ Tensoren sind. Wir behaupten aber auch folgendes. Wenn $A_{\mu\nu} B^{\mu\nu}$ für jede Wahl des Tensors $B^{\mu\nu}$ eine Invariante ist, so hat $A_{\mu\nu}$ Tensorcharakter.

　　　Beweis. Es ist nach Voraussetzung für eine beliebige Substitution

$$A_{\sigma\tau}' B^{\sigma\tau'} = A_{\mu\nu} B^{\mu\nu}$$

Nach der Umkehrung von (9) ist aber

$$B^{\mu\nu} = \frac{\partial x_\mu}{\partial x_\sigma'} \frac{\partial x_\nu}{\partial x_\tau'} B^{\sigma\tau'}$$

Dies eingesetzt in obige Gleichung liefert

$$\left( A_{\sigma\tau}' - \frac{\partial x_\mu}{\partial x_\sigma'} \frac{\partial x_\nu}{\partial x_\tau'} A_{\mu\nu} \right) B^{\sigma\tau'} = 0.$$

Dies kann bei beliebiger Wahl von $B^{\sigma\tau'}$ nur dann erfüllt sein, wenn die Klammer verschwindet, woraus mit Rücksicht auf (11) die Behauptung folgt.

Dieser Satz gilt entsprechend auch für Tensoren beliebigen Art Ranges und Charakters; der Beweis ist stets analog zu führen.

　　　Ebenso gilt der Satz: Wenn $A_{\mu\nu} B^\nu$ bei beliebiger Wahl des Vierervektors $B^\nu$ ein Tensor ist, so ist $A_{\mu\nu}$ ein Tensor. Der Beweis ist ganz analog dem soeben gegebenen.

　　　Der Satz lässt sich ebenso beweisen in der Form: Sind $B^\mu$ und $C^\nu$ beliebige Vektoren, und ist bei jeder Wahl derselben das innere Produkt

$$A_{\mu\nu} B^\mu C^\nu$$

eine Skalar, so ist $A_{\mu\nu}$ ein kovarianter Tensor. Dieser letztere Satz gilt auch dann noch, wenn nur bekannt ist, dass nur die speziellere Aussage zutrifft, dass bei beliebiger Wahl des Vierervektors $B^\mu$ das skalare Produkt

$$A_{\mu\nu} B^\mu B^\nu$$

## What was Einstein's strategy in constructing a gravitational field equation?

We have already emphasized (p. 57) that general covariance requires that the terms in the mathematical equations representing the laws of physics be tensors. Therefore, it is important to understand the rules of formation of tensors and the tensor character of mathematical objects in the context of the physical applications to which Einstein is leading. Thus, he continues to present the rules of the formation of tensors by multiplication of two tensors—specifically, by "external multiplication," which leads to a new tensor whose rank is the sum of the ranks of the two tensors, and by "mixed multiplication," which is a combination of an "external multiplication" and contraction of an upper index of one of the tensors with a lower one of the other.

Einstein takes great pains in elaborating and explaining all the mathematical instruments needed to express his physical ideas. He correctly assumed that these instruments were unfamiliar to contemporary physicists. Today, these instruments are standard in any introductory texts on general relativity and therefore need no further commentary here. Instead, let us look back at the formative years when Einstein was considering his plan of action.

The principles, listed on the preceding page, could serve as starting guidelines for the construction of a field equation or as validity criteria of such an equation. The right-hand side of the field equation represents the source of the field, and the left-hand side describes by means of a specific mathematical procedure—a so-called differential operator—how that source generates the field. In this context, two heuristic strategies can be identified in Einstein's search for the final field equation. One of them has been referred to as the "physical strategy," in which Einstein started with an object representing the left-hand side of the equation, which gave the correct law of gravitation in the classical Newtonian limit, modified it to conserve energy and momentum, and finally checked the degree of covariance of the obtained field equation. Thus, this strategy began with an attempt to satisfy the requirements of the correspondence and conservation principles. In the complementary "mathematical strategy," based on the newly acquired knowledge of Riemannian geometry, Einstein began with appropriate candidates for the left-hand side of the field equation, satisfying the general relativity principle, and then checked the compatibility with the physical requirements of the correspondence and conservation principles.

Einstein's double strategy arose from the different roles of the heuristic requirements of the principles of general relativity, correspondence, and energy-momentum conservation in the structure of the theory.

Actually, Einstein's efforts in the years 1912–1915 can be described as an interplay between these two complementary heuristic strategies.

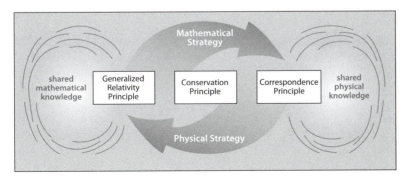

ein Skalar ist, falls man ausserdem weiss, dass $A_{\mu\nu}$ der Symmetrie-
bedingung $A_{\mu\nu} = A_{\nu\mu}$ genügt. Denn auf dem vorhin angegebenen Wege
beweist man den Tensorcharakter von $(A_{\mu\nu} + A_{\nu\mu})$, woraus dann wegen
der Symmetrie-Eigenschaft von $A$ der Tensorcharakter von $A_{\mu\nu}$ selbst
folgt. Auch dieser Satz lässt sich leicht verallgemeinern auf den Fall
kovarianter und kontravarianter Tensoren beliebigen Ranges.

Endlich folgt aus dem Bewiesenen der ebenfalls auf beliebige
Tensoren zu verallgemeinernde Satz: Wenn das innere Produkt die
Grössen $A_{\mu\nu} B^{\nu}$ bei beliebiger Wahl des Vierervektors $B^{\nu}$ einen Tensor
ersten Ranges bilden, so ist $A_{\mu\nu}$ ein Tensor zweiten Ranges. Ist
nämlich $C^{\mu}$ ein beliebiger Vierervektor, so ist wegen des Tensorcharakters $A_{\mu\nu} B^{\nu}$
das innere Produkt $A_{\mu\nu} C^{\mu} B^{\nu}$ bei beliebiger Wahl der beiden Vierervektoren $C^{\mu}$ und $B^{\nu}$ ein Skalar, woraus die Behauptung folgt.

§ 8. Einiges über den Fundamentaltensor der $g_{\mu\nu}$.

In dem Ausdruck des Quadrates des Linienelementes

$$ds^2 = g_{\mu\nu}\, dx_\mu\, dx_\nu$$

spielt $dx_\mu$ die Rolle eines beliebig wählbaren kontravarianten Vektors.
Da ferner $g_{\mu\nu} = g_{\nu\mu}$, so folgt nach den Betrachtungen des letzten § hieraus, dass $g_{\mu\nu}$ ein kovarianter Tensor zweiten Ranges ist. Wir nennen ihn "Fundamental-
tensor". Im Folgenden leiten wir einige Eigenschaften dieses Tensors ab,
die zwar jedem Tensor zweiten Ranges eigen sind; aber die besondere Rolle
welche des Fundamentaltensors in unserer Theorie, welche in der Besonder-
heit der Gravitationswirkungen ihren physikalischen Grund hat, bringt es
mit sich, dass die zu entwickelnden Relationen bei dem Fundamentaltensor für uns von Bedeutung sind.

Der kontravariante Fundamentaltensor. Bildet man in dem Determinantenschema
der $g_{\mu\nu}$ zu jedem $g_{\mu\nu}$ die Unterdeterminante und dividiert diese durch
die Determinante $g = |g_{\mu\nu}|$ der $g_{\mu\nu}$, so erhält man gewisse Grössen $g^{\mu\nu}(= g^{\nu\mu})$,
von denen wir beweisen wollen, dass sie einen kontravarianten
Tensor bilden.

Nach einem bekannten Determinantensatze ist

$$g_{\mu\sigma}\, g^{\nu\sigma} = \delta_\mu^\nu, \quad \ldots \ldots (16)$$

wobei das Zeichen $\delta_\mu^\nu$ 1 oder 0 bedeutet, je nachdem $\mu = \nu$
oder $\mu \neq \nu$ ist. Statt des obigen Ausdrucks für $ds^2$ können wir auch

$$g_{\mu\sigma}\, \delta_\nu^\sigma\, dx_\mu\, dx_\nu$$

oder nach (16) auch

$$g_{\mu\sigma}\, g_{\nu\tau}\, g^{\sigma\tau}\, dx_\mu\, dx_\nu$$

## Why is the metric tensor so fundamental?

Readers who have followed Einstein's exposition will remember how the measurement of distances in Euclidean space is generalized to spaces of more complicated geometries. In a four-dimensional spactime 10 functions are required to this end. They were introduced on page 52 [8] in the context of the calculation of the line element in curvilinear coordinates. Together these functions constitute a symmetric covariant tensor of rank 2, which Einstein refers to as the *fundamental tensor*, known today as the *metric tensor*.

In section 8, Einstein begins to explore the properties of this tensor, which allows calculation of the line element $ds$, namely, the distance between two neighboring points in spacetime, in terms of the projections $dx_\mu$ on the basis vectors, determined by the chosen coordinate system (the unnumbered equation in the middle of the page, which appeared previously on p. 8). In this equation, $g_{\mu\nu}$ is a covariant tensor, and $dx_\mu$ plays the role of a contravariant vector (in spite of the lower index). In Euclidean space, with Cartesian coordinates, this tensor reduces to the unit matrix $(g_{11} = g_{22} = g_{33} = g_{44} = 1)$. In the Minkowski space of special relativity, $g_{11} = g_{22} = g_{33} = -1$, $g_{44} = 1$, and all the off-diagonal elements are zero (eq. 4 on p. 52 [8]). In Euclidean space with curvilinear coordinates, or in curved spacetime, related to accelerated reference frames or to gravitational fields, the metric in general has 10 independent components that are functions of space (spacetime). The matrix representing $g_{\mu\nu}$ has an inverse matrix, namely, a matrix that when multiplied by $g_{\mu\nu}$ gives the unit matrix (eq. 16). This inverse matrix is the contravariant metric tensor $g^{\mu\nu}$.

The metric tensor is the common element of two mathematical traditions. It appeared in the expression for the line element in the tradition of differential geometry as well as in the tradition of vector analysis of Euclidean space described in terms of curvilinear coordinates. The development of vector and tensor calculus was closely associated with the interaction between physics and mathematics in the nineteenth and twentieth centuries. Although the direction of forces already played an important role in mechanics, it was only in the context of the development of electrodynamics in the late nineteenth century that the vector concept gained prominence because of its role in describing the directional properties of the electromagnetic field. At about the same time, the tensor concept emerged in the context of crystallography to describe the symmetry properties of crystals. Einstein and Grossmann, through their work on the theory of general relativity, combined this tradition of vector and tensor analysis with the work of Riemann, Christoffel, Ricci, and Levi-Civita on differential geometry and on the theory of invariants. In the elaboration of their mathematical framework, Einstein and Grossmann created a tensor concept that was much more general than the mathematical objects that had been used first in crystallography and then in electrodynamics and special relativity.

schreiben. Nun ~~aber~~ bilden aber nach den Multiplikationsregeln des vorigen § die Grössen

$$d\xi_\sigma = g_{\mu\sigma}\, dx_\mu$$

~~und~~ $$d\xi_\tau = g_{\rho\tau}\, dx_\rho$$

einen (Kovarianten ~~Vierervektor~~, und zwar (wegen der willkürlichen Wählbarkeit der $dx_\mu$) ~~beliebig~~ einen beliebig wählbaren Vierervektor. Indem wir ihn in unseren Ausdruck einführen, erhalten wir

$$ds^2 = g^{\sigma\tau}\, d\xi_\sigma\, d\xi_\tau$$

Da dies bei beliebiger Wahl des Vektors $d\xi_\sigma$ ein Skalar ist, und $g^{\sigma\tau}$ nach seiner Definition in den Indizes $\sigma$ und $\tau$ symmetrisch ist, folgt aus den Ergebnissen des vorigen §, dass $g^{\sigma\tau}$ ein kontravarianter Tensor ist. ~~Ebenso folgt noch aus (16),~~ folgt noch, dass auch $\delta_\sigma^\tau$ ein Tensor ist, den wir ~~als~~ den gemischten ~~Fundamentaltensor~~ ~~bezeichnen~~ nennen können.

Determinante des Fundamentaltensors. Nach dem Multiplikationssatz der Determinanten ist

$$\left| g_{\mu\alpha}\, g^{\alpha\nu} \right| = \left| g_{\mu\alpha} \right| \left| g^{\alpha\nu} \right|.$$

Andererseits ist

$$\left| g_{\mu\alpha}\, g^{\alpha\nu} \right| = \left| \delta_\mu^\nu \right| = 1$$

Also folgt

$$\left| g_{\mu\nu} \right| \left| g^{\mu\nu} \right| = 1 \cdots \cdots (17)$$

Invariante des Volumens. Wir suchen zuerst das Transformationsgesetz der Determinante $g = \left| g_{\mu\nu} \right|$. Gemäss (11) ist

$$g' = \left| \frac{\partial x_\mu}{\partial x_\sigma'} \frac{\partial x_\nu}{\partial x_\tau'} g_{\mu\nu} \right|$$

Hieraus folgt durch zweimalige Anwendung des Multiplikationssatzes der Determinanten

$$g' = \left| \frac{\partial x_\mu}{\partial x_\sigma'} \right| \left| \frac{\partial x_\nu}{\partial x_\tau'} \right| \left| g_{\mu\nu} \right| = \left| \frac{\partial x_\mu}{\partial x_\sigma'} \right|^2 g$$

oder

$$\sqrt{g'} = \left| \frac{\partial x_\mu}{\partial x_\sigma'} \right| \sqrt{g}.$$

Andererseits ist das Gesetz der Transformation des Volumelementes

$$d\tau = \int dx_1\, dx_2\, dx_3\, dx_4$$

nach dem bekannten Jakob'schen Satze

$$d\tau' = \left| \frac{\partial x_\sigma}{\partial x_\mu} \right| d\tau$$

### Why is the Zurich Notebook a unique document in the history of physics?

Einstein and Grossmann first explored the metric tensor in the Zurich Notebook. It is there that we find for the first time, in Einstein's handwriting, the expression for the line element $ds$ (first he denoted the metric tensor with a capital $G$ but soon adopted the notation with a lowercase $g$, which he used consistently after that).

$$ds^2 = \sum G_{\mu\nu}\, dx_\nu\, dx_\mu$$

© Hebrew University

After returning to Zurich from Prague in August 1912, Einstein started his collaboration with Grossmann to search for a gravitational field equation. Einstein documented his efforts during the winter of 1912/13 in a notebook, known as the Zurich Notebook, in which he filled whole pages with formulas and calculations but hardly any explanatory text. The Zurich Notebook is a unique document in the history of science, because it sheds light on the intricate process of a scientific discovery connected with a profound transformation of knowledge. The notebook allowed historians of science to decipher some of the blind alleys Einstein pursued and which even caused him to temporally abandon the goal of general covariance. The document shows how Einstein alternated between a mathematical strategy, starting from the Riemann tensor, and a physical strategy, starting from the classical equation for Newtonian gravity (the Poisson equation). In fact, he must have hoped that the two strategies would converge, as that would mean he had found a theory that combined his insights from the equivalence principle with the requirement of the Newtonian limit. Einstein was already very close to the solution of the problem at the end of 1912, but at the time the language of the new theory was not yet sufficiently mature to articulate how all these requirements could be reconciled.

### How are volumes measured in curved spacetime?

Einstein discusses on this page two related mathematical concepts: the "determinant of the fundamental tensor" and the "volume element." The determinant of a matrix is a number that can be calculated from its elements. The determinant of a diagonal matrix (with all the off-diagonal terms equal to zero) is just the product of the diagonal terms. Thus, the determinant of a unit matrix is equal to 1. The determinant of the matrix representing the metric in special relativity (Minkowski spacetime, eq. 4) is equal to −1. The determinant of the product of two matrices is the product of their determinants. The determinant corresponding to a covariant tensor is equal to the reciprocal of the determinant of its contravariant version.

© Hebrew University

To be able to formulate integrals in a curved spacetime one needs a volume measure, just as in three-dimensional Euclidean space, where one has the volume element $dx_1 dx_2 dx_3$. In a Riemannian space, the natural volume element $d\tau$ is given in terms of the coordinate differentials multiplied by the square root of the absolute value of the determinant of the metric. This page concludes with the transformation rules of the volume element between different coordinate systems, $x_\mu$ and $x'_\mu$.

(18)

Durch Multiplikation der beiden letzten Gleichungen erhält man

$$\sqrt{g'}\, d\tau' = \sqrt{g}\, d\tau \quad \ldots \ (18)$$

Statt $\sqrt{-g}$ wird im folgenden die Grösse $\sqrt{-g}$ eingeführt, welche wegen des hyperbolischen Charakters des Zeitraumlichen Kontinuums stets einen reellen Wert hat. Die Invariante $\sqrt{-g}\, d\tau$ ist gleich der Grösse des im „lokalen Bezugssystem" mit starren Massstäben und Uhren gemessenen vierdimensionalen Volumelementes.

Bemerkung über den Charakter des raum-zeitlichen Kontinuums. Unsere Voraussetzung, dass im Unendlichkleinen stets die spezielle Relativitätstheorie gelte, bringt es mit sich, dass sich $ds^2$ immer gemäss (1) durch die reellen Grössen $dX_1 \ldots dX_4$ ausdrücken lässt. Nennen wir $d\tau_0$ das „natürliche" Volumelement $dX_1\, dX_2\, dX_3\, dX_4$, so ist also

$$d\tau_0 = \sqrt{-g}\, d\tau \quad \ldots \ (18a)$$

Soll an einer Stelle des vierdimensionalen Kontinuums $\sqrt{-g}$ verschwinden, so bedeutet dies, dass hier einem endlichen Koordinatenvolumen ein unendlich kleines „natürliches" Volumen entspreche. Dies möge nirgends der Fall sein. Dann kann $g$ sein Vorzeichen nicht ändern; wir werden annehmen, dass $g$ stets negativen Wert habe. Es ist dies eine Hypothese über die physikalische Natur des betrachteten Kontinuums und gleichzeitig eine Festsetzung über die Koordinatenwahl.

Ist aber $-g$ stets positiv und endlich, so liegt es nahe, die Koordinatenwahl a posteriori so zu treffen, dass diese Grösse gleich 1 wird. Wir werden später sehen, dass durch eine solche Beschränkung der Koordinatenwahl eine bedeutende Vereinfachung der Naturgesetze erzielt werden kann. Anstelle von (18) tritt dann (18a) tritt dann einfach

$$d\tau' = d\tau,$$

woraus mit Rücksicht auf Jakobis Satz folgt

$$\left| \frac{\partial x'_\sigma}{\partial x_\mu} \right| = 1 \quad \ldots \ldots \ (19)$$

Bei dieser Koordinatenwahl sind also nur Substitutionen der Koordinaten von der Determinante 1 zulässig.

Es wäre aber irrtümlich, zu glauben, dass dieser Schritt

### How can a convenient choice of coordinates simplify the theory?

The determinants of the metric $g$ and the volume element $d\tau$ themselves are generally not invariant under coordinate transformations, but a certain combination of them is (eq. 18). In special relativity, the volume element is invariant because the determinant of the metric is invariant. In general relativity, the close neighborhood of a point in spacetime can be approximated by the Minkowski metric, just like the surroundings of every point on Earth can be approximated by a flat surface. The curvature comes into play in moving away from the local neighborhood of that point. However, a set of coordinate transformations known as *unimodular transformations* exists (mathematically characterized by eq. 19), for which any volume element is an invariant. In the Minkowski spacetime of special relativity and in general relativity the determinant of the metric is always negative. Hence, the restriction to unimodular transformations implies that $-g = 1$.

The group of transformations that render the equations of general relativity invariant can be broken down into two subgroups: the group of unimodular transformations and the group of volume-changing transformations. It can even be shown that the unimodular group considered here by Einstein is not only a technical simplification (as he then believed) but actually plays a fundamental role in most physical and mathematical applications.

Until this point the discussion has been purely mathematical. Now Einstein introduces the physical motivation: "We shall see later that by such a restriction of the choice of coordinates it is possible to achieve an important simplification of the laws of nature." A few lines later and on the next page, he emphasizes: "But it would be erroneous to believe that this step indicates a partial abandonment of the general postulate of relativity. We do not ask 'What are the laws of nature which are covariant in face of all substitutions for which the determinant is unity?' but our question is, 'What are the generally covariant laws of nature?' It is not until we have formulated these that we simplify their expression by a particular choice of the system of reference." He will repeat and reemphasize this point (see pp. 91 [27] and 119 [40a]).

### What is the difference between a *coordinate condition* and a *coordinate restriction*?

The issue of special sets of coordinates accompanied Einstein in the different phases of his search for a relativistic theory of gravitation. To check whether a generally covariant field equation can be reduced to the Newtonian limit (Poisson equation), one has to impose a specific set of coordinates in which Newton's theory holds true. Such a choice of coordinates is called today a *coordinate condition* and does not impinge on the general covariance of the theory.

It is in principle conceivable that a theory would privilege certain reference frames, as does special relativity with inertial frames. Such a situation could be expressed by a *coordinate restriction*. At the beginning, Einstein did not know whether he would succeed in keeping his new theory of gravitation free from such restrictions. In particular, it seemed that the requirement of energy-momentum conservation would necessitate such a coordinate restriction. This restriction would then still have to be compatible with the possibility to choose a coordinate system in which the Newtonian limit could be realized. Einstein faced a complicated interrelation between different requirements, which he did not know then how to disentangle, and which first led him to the *Entwurf* theory (p. 83 [23]).

The restriction to unimodular coordinates adopted here should be viewed as a coordinate condition intended to simplify the derivation of the final theory.

(18)

einen partiellen Verzicht auf das allgemeine Relativitätspostulat
bedeutete. Wir fragen nicht: "Wie heissen die Naturgesetze, welche gegenüber
allen Transformationen von der Determinante 1 kovariant sind?" Sondern
wir fragen: "Wie heissen die allgemein kovarianten Naturgesetze?" Erst
nachdem wir diese aufgestellt haben vereinfachen wir ihren Ausdruck durch eine besondere Wahl des Bezugssystems.

Bildung neuer Tensoren vermittelst des Fundamentaltensors. Durch innere,
äussere und gemischte Multiplikation eines Tensors mit dem Fundamental
tensor lassen sich nicht entstehen Tensoren anderen Charakters und Ranges. Beispiele:

$$A^\mu = g^{\mu\sigma} A_\sigma$$
$$A = g_{\mu\nu} A^{\mu\nu}$$

Besonders sei auf folgende Bildungen hingewiesen

$$A^{\mu\nu} = g^{\mu\alpha} g^{\nu\beta} A_{\alpha\beta}$$
$$A_{\mu\nu} = g_{\mu\alpha} g_{\nu\beta} A^{\alpha\beta}$$

("Ergänzung")

("Ergänzung" des kovarianten bezw. kontravarianten Tensors) und

$$B_{\mu\nu} = g_{\mu\nu} g^{\alpha\beta} A_{\alpha\beta}.$$

Wir nennen $B_{\mu\nu}$ den reduzierten, zu $A_{\mu\nu}$ gehörigen reduzierten
Tensor. Analog

$$B^{\mu\nu} = g^{\mu\nu} g_{\alpha\beta} A^{\alpha\beta}.$$

Es sei bemerkt, dass $g^{\mu\nu}$ nichts anderes ist als die Ergänzung von
$g_{\mu\nu}$. Denn man hat

$$g^{\mu\alpha} g^{\nu\beta} g_{\alpha\beta} = g^{\mu\alpha} \delta_\alpha^\nu = g^{\mu\nu}.$$

§9. Gleichung der geodätischen Linie (bezw. der Punktbewegung).
Da das "Linienelement" $ds$ eine physikalisch vollkommen definierte
Grösse ist, kann man zwischen zwei Punkten $P_1$ und $P_2$ des verdimen-
sionalen Kontinuums gezogenen Linie fragen, für welche $\int ds$ ein Ex-
tremum ist (Geodätische Linie). Ihre Gleichung ist

$$\delta \left\{ \int_{P_1}^{P_2} ds \right\} = 0 \quad \cdots \cdots \cdot (20)$$

Aus dieser Gleichung findet man in bekannter Weise durch Ausführung der Variation
vier totale Differentialgleichungen, welche diese geodätische Linie bestimmen;
auch diese Ableitung soll der Vollständigkeit halber hier Platz finden. Es
sei $\lambda$ eine Funktion der Koordinaten $x_\nu$; diese definiert
eine Schar von Flächen, jede welche die gesuchte geodätische
Linie sowie alle ihr unendlich benachbarten Linien durch die Punkte $P_1$ und $P_2$

### What is the meaning of a "straight line" in curved space, and how does a particle move under the influence of gravitation?

Einstein concludes this section on some aspects of the metric tensor by showing how this tensor may be used to form new tensors. An outer (p. 58 [11]), inner or mixed (p. 62 [13]) multiplication of a tensor by the metric tensor produces tensors of different character and rank. After a tedious exposition of a variety of properties of the metric tensor, Einstein had reached a point where he could introduce a new physically important concept.

In section 9, Einstein introduces the concept of the geodetic line and derives the mathematical equation satisfied by the points along this line.

Gaussian geometry is the study of curves and surfaces in three-dimensional Euclidean space. Geodetic lines on a curved surface are the lines of shortest distance between two points. For example, the shortest path between two points on a sphere is a section of the great circle passing through these points. This definition applies also to lines in a curved space of any dimension, except that it is not possible to visualize their shape in higher dimensions; they are described within the mathematical formalism of Riemannian geometry.

In the Zurich Notebook, Einstein derived a well-known result of classical mechanics: a particle constrained to move on a curved surface, without the influence of external forces, will move between two points along the geodetic line connecting these points. The same principle applies to the free motion of a particle in spacetime, where the effect of gravitation is reflected in the curvature of spacetime. Curiously, however, the geodetic line in a curved space, as observed from an arbitrary accelerated frame of reference or produced by an arbitrary gravitational field, turns out to be the longest possible path between two given points in spacetime. This is a consequence of the peculiar mathematical properties of the spacetime metric. In any case, the geodetic can always be defined as the path of extremal (minimum or maximum) distance between two points in spacetime.

The geodetic line is not only a mathematical object; it is the trajectory of force-free motion of a particle in a gravitational field.

The trajectory of the geodetic line is defined in a given frame of reference by the spacetime coordinates of the points on the line. These coordinates satisfy a mathematical equation that is derived here by means of the "variational method." This method is expressed compactly by eq. (20). The integral symbol $\int$ is a sum of the line elements $ds$ between point $P_1$ and point $P_2$, namely, the length of the path between the two points. The letter $\delta$ in front of the integral is an infinitesimal variation of this length for different trajectories. The trajectory for which the variation is zero is the trajectory of shortest (or longest) length and therefore represents the geodetic line. This is the natural generalization of a straight line in Riemannian geometry of curved space.

Eduard asked his father, Albert Einstein, why he was so famous (1922). Einstein responded: "When a blind beetle crawls over the surface of a curved branch, it doesn't notice that the track it has covered is indeed curved. I was lucky enough to notice what the beetle didn't notice."

(19)

schneiden. Jede solche Kurve kann dann dadurch gegeben gedacht werden, dass ihre Koordinaten $x_\nu$ in Funktion von $\lambda$ ausgedrückt werden. Das Zeichen $\delta$ entspreche dem Übergang von einem Punkte der gesuchten geodätischen Linie zu denjenigen Punkte einer benachbarten Kurve, welcher zu dem nämlichen $\lambda$ gehört. Dann lässt sich (20) durch

$$\int_{\lambda_1}^{\lambda_2} \delta w \, d\lambda = 0 \qquad \left.\right\} (20a)$$

$$w^2 = g_{\mu\nu} \frac{dx_\mu}{d\lambda} \frac{dx_\nu}{d\lambda}$$

ersetzen. Da aber

$$\delta w = \frac{1}{w} \left\{ \frac{1}{2} \frac{\partial g_{\mu\nu}}{\partial x_\sigma} \frac{dx_\mu}{d\lambda} \frac{dx_\nu}{d\lambda} \delta x_\sigma + g_{\mu\nu} \frac{dx_\mu}{d\lambda} \delta \left( \frac{dx_\nu}{d\lambda} \right) \right\},$$

nach Einsetzen von $\delta w$ in (20a)

so erhält man mit Rücksicht darauf, dass

$$\delta \left( \frac{dx_\nu}{d\lambda} \right) = \frac{d \delta x_\nu}{d\lambda},$$

nach partieller Integration

$$\int_{\lambda_1}^{\lambda_2} d\lambda \, K_\sigma \, \delta x_\sigma = 0 \qquad \left.\right\} (20b)$$

$$K_\sigma = \frac{d}{d\lambda} \left\{ \frac{g_{\mu\nu}}{w} \frac{dx_\mu}{d\lambda} \right\} - \frac{1}{2w} \frac{\partial g_{\mu\nu}}{\partial x_\sigma} \frac{dx_\mu}{d\lambda} \frac{dx_\nu}{d\lambda}$$

Hieraus folgt wegen der freien Wählbarkeit der $\delta x_\sigma$ das Verschwinden der $K_\sigma$. Also sind

$$K_\sigma = 0 \quad \cdots \cdots (20c)$$

die Gleichungen der geodätischen Linie. Ist auf der betrachteten geodätischen Linie nicht $ds = 0$, so können wir als Parameter $\lambda$ die auf der geodätischen Linie gemessene "Bogenlänge" $s$ wählen. Dann wird $w = 1$, und man erhält anstelle von (20c)

$$g_{\mu\nu} \frac{d^2 x_\mu}{ds^2} + \frac{\partial g_{\mu\nu}}{\partial x_\sigma} \frac{dx_\sigma}{d\lambda} \frac{dx_\mu}{d\lambda} - \frac{1}{2} \frac{\partial g_{\mu\nu}}{\partial x_\sigma} \frac{dx_\mu}{d\lambda} \frac{dx_\nu}{d\lambda} = 0,$$

oder durch blosse Aenderung der Bezeichnungsweise

$$g_{\alpha\sigma} \frac{d^2 x_\alpha}{ds^2} + \left[ \begin{matrix} \mu\,\nu \\ \sigma \end{matrix} \right] \frac{dx_\mu}{ds} \frac{dx_\nu}{ds} = 0, \quad \cdots (20d)$$

nach Christoffel

wobei gesetzt ist

$$\left[ \begin{matrix} \mu\,\nu \\ \sigma \end{matrix} \right] = \frac{1}{2} \left( \frac{\partial g_{\mu\sigma}}{\partial x_\nu} + \frac{\partial g_{\nu\sigma}}{\partial x_\mu} - \frac{\partial g_{\mu\nu}}{\partial x_\sigma} \right). \quad \cdots (21)$$

## What is the geometric and physical significance of "Christoffel symbols"?

The derivation of the geodetic line equation by means of the variational method, described on the preceding page, leads to the definition of a mathematical object that plays a central role in tensor calculus (differential geometry)—the Christoffel symbol—defined at the bottom of the page (eq. 21), and a slightly different version thereof is given on the next page (eq. 23). The symbol describes what happens to vectors and tensors when they are moved along a line in a curved space. It is indispensable for tracing the path of geodetic lines, for calculating derivatives of tensors, and for characterizing the local properties of specific Riemannian and spacetime geometries.

   Why do we need another such fundamental object? We have already mentioned that in general relativity the gravitational potential is represented by the metric tensor. The Christoffel symbol, which is a combination of derivatives of the elements of the metric tensor, plays the role of the gravitational force field. We shall return to the Christoffel symbols in subsequent pages, both in the context of mathematical concepts in differential geometry and of their physical significance in general relativity. Let us now draw attention to one more point. We have emphasized that mathematical equations representing physical laws have to be expressed as equations between tensors, yet the Christoffel symbols are not tensors. Thus, they will never appear alone in such equations.

## What was Einstein's "fatal prejudice" in the early identification of the gravitational field components?

In classical physics, to every point in space around a distribution of massive bodies one can assign a number, the gravitational potential, which measures the gravitational energy of a particle of unit mass at that point. Particles that are free to move will move from points of higher to points of lower potential. At every point in space, a particle that is free to move will move in the direction of the gravitational force vector at that point with an acceleration determined by the strength of the field. The components of the gravitational field are the local changes of the gravitational potential along the direction of the space coordinates (the derivatives with respect to the space coordinates).

   As mentioned before (p. 55), in general relativity, the single gravitational potential function of Newtonian physics is replaced by 10 functions of spacetime coordinates, which are the 10 independent components of the metric tensor $g_{\mu\nu}$. The gravitational field components are again determined by derivatives of the components of the gravitational potential. To assure the general covariance of the theory, these derivatives have to be calculated by the rules of covariant differentiation (p. 79 [21]). This process leads to the identification of the gravitational field components as the Christoffel symbols.

   The realization that the Christoffel symbols, and not simple derivatives of the gravitational potentials $g_{\mu\nu}$, are the gravitational field components was the key element in the final phase of the derivation of the general theory of relativity in November 1915. Previously, in 1913, Einstein had associated a different mathematical expression with the gravitational field, which led to the *Entwurf* theory (p. 83). On November 4, 1915, in a presentation to the Prussian Academy, he confessed that this was "a fatal prejudice."

Multipliziert man endlich (20h.) mit $g^{\sigma\tau}$ (äussere Multiplikation bezüglich $\tau$, innere bezüglich $\sigma$), so erhält man schliesslich als endgültige Form der Gleichung der geodätischen Linie

$$\frac{d^2 x_\tau}{ds^2} + \left\{\begin{matrix}\mu\nu\\\tau\end{matrix}\right\} \frac{dx_\mu}{ds}\frac{dx_\nu}{ds} = 0. \quad \cdots (22)$$

Hiebei ist nach Christoffel gesetzt

$$\left\{\begin{matrix}\mu\nu\\\tau\end{matrix}\right\} = g^{\tau\alpha}\left[\begin{matrix}\mu\nu\\\alpha\end{matrix}\right] \quad \cdots \cdots (23)$$

§10. Die Bildung von Tensoren durch Differentiation.

Gestützt auf die Gleichung der geodätischen Linie können wir nun leicht die Gesetze ableiten, nach welchen durch Differentiation aus Tensoren neue Tensoren gebildet werden können. Dadurch werden wir erst in den Stand gesetzt, allgemein kovariante Differentialgleichungen aufzustellen. Wir erreichen das Ziel durch folgenden einfachen Satz.

Ist in unserem Kontinuum eine Kurve gegeben, deren Punkte durch die Bogendistanz $s$ von einem festen Punkt auf der Kurve charakterisiert sind, ist ferner $\varphi$ eine invariante Raumfunktion, so ist auch $\frac{d\varphi}{ds}$ eine Invariante. Der Beweis liegt darin, dass sowohl $d\varphi$ als auch $ds$ Invariante sind.

Da $\frac{d\varphi}{ds} = \frac{\partial\varphi}{\partial x_\mu}\frac{dx_\mu}{ds}$, so ist auch

$$\psi = \frac{\partial\varphi}{\partial x_\mu}\frac{dx_\mu}{ds}$$

eine Invariante, und zwar für alle Kurven, die von einem Punkte des Kontinuums ausgehen, das heisst für beliebige Wahl des Vektors der $dx_\mu$. Daraus folgt unmittelbar, dass

$$A_\mu = \frac{\partial\varphi}{\partial x_\mu} \quad \cdots \cdots (24)$$

ein kovarianter Vierervektor ist (Gradient von $\varphi$).

Nach unserem Satze ist ebenso $\chi = \frac{d\psi}{ds}$ eine Invariante. Durch Einsetzen von $\psi$ erhalten wir zunächst

$$\chi = \frac{\partial^2\varphi}{\partial x_\mu \partial x_\nu}\frac{dx_\mu}{ds}\frac{dx_\nu}{ds} + \frac{\partial\varphi}{\partial x_\mu}\frac{d^2 x_\mu}{ds^2}$$

Hieraus lässt sich zunächst die Existenz einer Kovariante nicht ableiten. Setzen wir nun aber fest, dass die Kurve, auf welcher wir differenziert haben, eine geodätische Kurve sei, so erhalten wir nach (22) durch Ersetzen von $\frac{d^2 x_\nu}{ds^2}$

$$\chi = \left\{\frac{\partial^2\varphi}{\partial x_\mu \partial x_\nu} - \left\{\begin{matrix}\mu\nu\\\tau\end{matrix}\right\}\frac{\partial\varphi}{\partial x_\tau}\right\}\frac{dx_\mu}{ds}\frac{dx_\nu}{ds}.$$

Aus der Vertauschbarkeit der Differentiationen nach $\mu$ und $\nu$ und daraus, dass gemäss (23) und (21) die Klammer $\left\{\begin{matrix}\mu\nu\\\tau\end{matrix}\right\}$ bezüglich $\mu$ und $\nu$ symmetrisch ist, folgt, dass der Klammerausdruck in $\mu$ und $\nu$ symmetrisch ist.

## The geodetic line as the "straightest" possible line and its relation to the concept of "affine connection"

On page 18, the geodetic line is defined as the line of extremal, either shortest or longest, distance between two points in spacetime, and its equation is derived by means of the variational method. The result of this calculation is given in eq. (22). We shall meet this equation again in a slightly different notation on page 92 [28]. There it is presented as the equation of motion of a particle in a gravitational field.

Under certain conditions, the geodetic line can also be characterized as the line for which the tangent vector remains parallel to itself when moved from point to point along the line (see illustration on preceding page). The tangent vector is a unit vector along the tangent direction at a given point on the line. Intuitively, this requirement means that the geodetic line is the "straightest" possible line between two points. This definition will turn out to be important for understanding the concept and process of differentiation in a curved space.

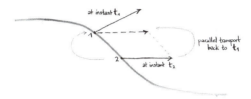

This definition of a geodetic line relies on the concept of a *parallel displacement* or *parallel transport* of a vector. To apply this concept in Riemannian geometry, one has to understand how a geometric situation at one point can be compared with that at another point. In this context, mathematicians in the nineteenth century began to explore the notion of *connection*, which describes how geometric data are consistently transported along specific curves. The most elementary type of connection and the most relevant for our discussion is the *affine connection*, which specifies how vectors are transported in parallel, along a curve, from one point to another. The affine connection is intimately related to the derivative of a vector in a certain direction, namely, to the question, how is a vector changed by an infinitesimal transport in a given direction?

Historically, the infinitesimal perspective of connections in Riemannian geometry began with Christoffel and was studied in greater detail in the beginning of the twentieth century by Levi-Civita and Ricci. They established a relation between infinitesimal connections, discussed by Christoffel, and the notion of parallel transport. Levi-Civita used the notion of parallel transport to clarify and illustrate the concept of covariant differentiation. In this context, the affine connection is also known as the *Levi-Civita connection* and is identified with the Christoffel symbol itself.

The next section is devoted to the formation of tensors from given tensors by differentiation. This is an important topic because the laws of physics are represented by differential equations, and these equations have to be generally covariant. Einstein was aware that these laws had already been derived by mathematicians, but he preferred to do it his own way. In a previous review article, "Foundation of General Relativity," submitted to the Prussian Royal Academy in October 1914, he wrote: "The laws of these differential expressions have already been given by Christoffel, Ricci and Levi-Civita. I give here a particularly simple derivation of this, which appears to be new." The derivation there is identical to the one presented here.

(21)

Da man von einem Punkt des Kontinuums aus in beliebiger Richtung eine geodätische Linie ziehen kann, $\frac{dx_\mu}{ds}$ also ein Vierervektor mit frei wählbarem Verhältnis der Komponenten ist, folgt nach den Ergebnissen des § 7, dass

$$A_{\mu\nu} = \frac{\partial^2 \varphi}{\partial x_\mu \partial x_\nu} - \left\{{\mu\ \nu \atop \tau}\right\}\frac{\partial\varphi}{\partial x_\tau} \quad \cdots \cdots (25)$$

ein kovarianter Tensor zweiten Ranges ist. Wir haben also das Ergebnis gewonnen: Aus dem kovarianten Viervektor Tensor ersten Ranges $A_\mu = \frac{\partial\varphi}{\partial x_\mu}$ können wir durch Differentiation einen kovarianten Tensor zweiten Ranges

$$A_{\mu\nu} = \frac{\partial A_\mu}{\partial x_\nu} - \left\{{\mu\ \nu \atop \tau}\right\} A_\tau \quad \cdots \cdots (26)$$

bilden. Wir nennen den Tensor $A_{\mu\nu}$ die „Erweiterung" des Tensors $A_\mu$. Zunächst können wir leicht zeigen, dass diese Bildung auch dann auf einen Tensor führt, wenn der Vektor $A_\mu$ nicht als ein Gradient darstellbar ist. Um dies einzusehen, bemerken wir zunächst, dass $\psi\frac{\partial\varphi}{\partial x_\mu}$ ein kovarianter Vierervektor ist, wenn $\psi$ und $\varphi$ Skalare sind. Dies ist auch der Fall für eine aus vier solchen Gliedern bestehende Summe

$$S_\mu = \psi^{(1)}\frac{\partial\varphi^{(1)}}{\partial x_\mu} + \cdots + \psi^{(4)}\frac{\partial\varphi^{(4)}}{\partial x_\mu},$$

falls $\psi^{(1)}\varphi^{(1)}, \ldots \psi^{(4)}\varphi^{(4)}$ Skalare sind. Nun ist aber klar, dass sich jeder kovariante Vierervektor in der Form $S_\mu$ darstellen lässt. Hat nämlich $A_\mu$ ein Vierervektor, dessen Komponenten beliebige gegebene Funktionen (bezüglich des gewählten Koordinatensystems) der $x_\nu$ sind, so hat man nur zu setzen

$$\psi^{(1)} = A_1 \qquad \varphi^{(1)} = x_1$$
$$\psi^{(2)} = A_2 \qquad \varphi^{(2)} = x_2$$
$$\psi^{(3)} = A_3 \qquad \varphi^{(3)} = x_3$$
$$\psi^{(4)} = A_4 \qquad \varphi^{(4)} = x_4,$$

um zu erreichen, dass $S_\mu$ gleich $A_\mu$ wird.

Um daher zu beweisen, dass (26) $A_{\mu\nu}$ ein Tensor ist, wenn auf der rechten Seite für $A_\mu$ ein beliebiger kovarianter Vierervektor eingesetzt wird, brauchen wir nur zu zeigen, dass dies für den Vierervektor $S_\mu$ zutrifft. Für letzteres ist es aber, weil die rechte Seite von (26) linear und homogen bezüglich $A_\tau$ und $\frac{\partial A_\mu}{\partial x_\nu}$ ist, hinreichend, den Nachweis für den Fall

$$A_\mu = \psi\frac{\partial\varphi}{\partial x_\mu}$$

zu führen. Es ist nun die mit $\psi$ multiplizierte rechte Seite von (25)

$$\psi\frac{\partial\varphi}{\partial x_\mu \partial x_\nu} - \left\{{\mu\ \nu \atop \tau}\right\}\frac{\partial\varphi}{\partial x_\tau}$$

Tensorcharakter. Ebenso

### How do tensors change between neighboring points, or how can one produce new tensors from given tensors by differentiation?

Einstein's derivation begins with the differentiation of a scalar function (previous page). He shows that the four partial derivatives, with respect to the coordinates, form a (covariant) vector (eq. 24) and fully describe how this scalar function varies between neighboring points in space. He reaches this conclusion by exploring the change of the scalar function along a curve, on which the points are parameterized by their distance $s$ from a fixed point on the curve.

> The next question is, how does this vector, or any (covariant) vector, vary from point to point? The first thought could be that such a variation is characterized by the partial derivatives of the four components of the vector. It turns out, however, that the 16 parameters obtained by this procedure do not form a tensor of rank 2. Moreover, these parameters are not sufficient to describe the change of a vector in curvilinear coordinates, in which there are two contributions to the change of a vector. First, its direction and/or size may vary. Second, the components of the vector vary because the basis vectors, tangent to the coordinate lines, which define these components, change from point to point. The latter effect is present even in the case of a constant vector.

Einstein shows how one can obtain a tensor by differentiation of the vector in eq. (24). To this end, he assumes that the curve along which the differentiation is performed is a geodetic curve, so that he can use the equation of a geodetic curve and derive the tensor given in eq. (26). Einstein refers to this tensor, $A_{\mu\nu}$, as the "extension" of the vector $A_\mu$. Today, it is called the *covariant derivative* of the vector $A_\mu$, and the procedure leading to it is called *covariant differentiation*. In modern understanding, it crucially involves the concept of parallel transport of a vector, making it possible to compare vectors at different points along a curve (see illustration).

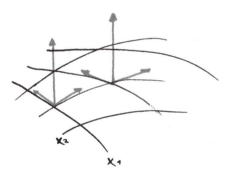

The two contributions to the variation of a vector between neighboring points in space, described in the previous paragraph, are represented by the two terms on the right-hand side of eq. (26). The second term, which is a consequence of the curvilinear nature of the coordinate system, involves the Christoffel symbol. Neither of the two terms is a tensor, but their difference (sum) is a tensor of rank 2.

Einstein first established this result for a vector derived from a scalar function of coordinates (eq. 24) and then generalized it to any covariant vector. He did this by showing that any (covariant) vector may be represented as a sum of four such vectors derived from scalar functions by differentiation.

(22)

ist

$$\frac{\partial \psi}{\partial x_\mu}\,\frac{\partial \psi}{\partial x_\nu}$$

ein Tensor (äusseres Produkt zweier Vierervektoren). Durch Addition folgt der Tensorcharakter von

$$\frac{\partial}{\partial x_\nu}\left(\psi\frac{\partial \psi}{\partial x_\mu}\right) - \left\{{}^{\mu\ \nu}_{\ \tau}\right\}\left(\psi\frac{\partial \psi}{\partial x_\tau}\right).$$

Damit ist, wie ein Blick auf (26) lehrt, der verlangte Nachweis für den Vierervektor $\psi\frac{\partial \psi}{\partial x_\mu}$, und daher nach dem vorhin Bewiesenen für jeden beliebigen Vierervektor $A_\mu$ geführt. —

Mit Hilfe der "Erweiterung" des Vierervektors kann man leicht die "Erweiterung" eines kovarianten Tensors beliebigen Ranges definieren; diese Bildung ist eine Verallgemeinerung der Erweiterung des Vierervektors. Wir beschränken uns auf die Aufstellung der Erweiterung des Tensors zweiten Ranges, da dieser das Bildungsgesetz bereits klar übersehen lässt.

Wie bereits bemerkt, lässt sich jeder kovariante Tensor zweiten Ranges darstellen als eine Summe von Tensoren vom Typus $A_\mu B_\nu$. Es wird deshalb genügen, den Ausdruck der Erweiterung für einen solchen speziellen Tensor abzuleiten. Nach (26) haben die Ausdrücke

$$\frac{\partial A_\mu}{\partial x_\sigma} - \left\{{}^{\sigma\ \mu}_{\ \tau}\right\}A_\tau$$

$$\frac{\partial B_\nu}{\partial x_\sigma} - \left\{{}^{\sigma\ \nu}_{\ \tau}\right\}B_\tau$$

Tensorcharakter. Durch äussere Multiplikation des ersten mit $B_\nu$, des zweiten mit $A_\mu$ erhält man einen Tensor dritten Ranges; deren Addition ergibt den Tensor dritten Ranges

$$A_{\mu\nu\sigma} = \frac{\partial A_{\mu\nu}}{\partial x_\sigma} - \left\{{}^{\sigma\ \mu}_{\ \tau}\right\}A_{\tau\nu} - \left\{{}^{\sigma\ \nu}_{\ \tau}\right\}A_{\mu\tau}, \quad \ldots (27)$$

wobei $A_{\mu\nu} = A_\mu B_\nu$ gesetzt ist. Da die rechte Seite von (27) linear und homogen ist bezüglich der $A_{\mu\nu}$ und deren erster Ableitungen, führt dies Bildungsgesetz nicht nur bei einem Tensor von Typus $A_\mu B_\nu$ sondern auch bei einer Summe solcher Tensoren, d.h. bei einem beliebigen kovarianten Tensor zweiten Ranges zu einem Tensor. Wir nennen $A_{\mu\nu\sigma}$ die Erweiterung des Tensors $A_{\mu\nu}$.

Es ist klar, dass (26) und ( ) nur spezielle Fälle von (27) sind (Erweiterung des Tensors ersten bezw. nullten Ranges). Überhaupt

x Durch äussere Multiplikation der Vektoren mit den Komponenten (beliebig gegeben) $A_{11}, A_{12}, A_{13}, A_{14}$ bezw. 1, 0, 0, 0 entsteht ein Tensor mit den Komponenten $\begin{smallmatrix}A_{11} & A_{12} & A_{13} & A_{14}\\ 0 & 0 & 0 & 0\\ 0 & 0 & 0 & 0\\ 0 & 0 & 0 & 0\end{smallmatrix}$. Nach Addition (Komponenten) von vier Tensoren von diesem Typus erhält man den Tensor $A_{\mu\nu}$ mit beliebig vorgeschriebenen)

Einstein generalizes the procedure of covariant differentiation to covariant tensors of rank 2. He derives the result for a tensor, which is produced by outer multiplication of two covariant vectors, because every covariant tensor of rank 2 can be represented as a sum of four such tensors. In the footnote, he demonstrates this statement. For a tensor of rank 2, one has to take into account the effect of the change of the basis vectors between two neighboring points in curvilinear coordinates (see explanation on previous page) on the two indices. Therefore, the covariant derivative ("extension") of a covariant tensor of rank 2 (a covariant tensor of rank 3) has two terms with Christoffel symbols (eq. 27).

### What is the geometric context of Einstein's mathematical formulation of general relativity?

The mathematical framework of Einstein's general relativity emerged from the absolute differential calculus of Christoffel, Ricci, and Levi-Civita. This framework was built around the concept of differential invariants. Its relation to differential geometry and its geometric interpretation became more prominent only after the establishment of Einstein's theory. Hermann Weyl, in particular, clarified the geometric interpretation of the Riemann-Christoffel curvature tensor and related it to the parallel displacement of a vector around a closed loop.

In his mathematical exposition, Einstein discussed the equation and meaning of the geodetic line and thus introduced a key element of non-Euclidean geometry that did not figure in the work of Ricci and Levi-Civita. He also realized early on that the four-dimensional spacetime of general relativity no longer fit the framework of Euclidean geometry when he considered the thought experiment of the rotating disk in 1912. However, he did not systematically introduce non-Euclidean geometry, nor did he interpret his own theory in terms of differential geometry. When he discusses the Riemann-Christoffel tensor, for instance, he does not even mention curvature. The geometrization of general relativity and the understanding of gravity as being due to the curvature of spacetime is a result of the further development and not a presupposition of Einstein's formulation of the theory.

In May 1921, Einstein gave a series of lectures on special and general relativity at Princeton University. There, unlike in the present manuscript, he acknowledged that covariant differential operations on tensors are most satisfactorily recognized by the method introduced by Levi-Civita and later used in the context of general relativity by Weyl. In a given vector field, the specific vector at point $P_1$ is shifted, parallel to itself, to a neighboring point $P_2$. The difference between the shifted vector and the field vector at $P_2$ may be regarded as the differential of the vector at $P_1$. Although Einstein did not use the term "affine connection," this is exactly what it is. This calculation naturally renders the Christoffel symbol (p. 78 [21]).

Toward the end of his life, Einstein summarized his earlier interpretation of his own theory. He stressed the role of Levi-Civita's notion of the displacement of vectors, which was developed only after the completion of general relativity, rather than the Riemannian concept of a metric as the appropriate mathematical representation of the conceptual key insight of general relativity: the role of "background independence":

> It is well known that around the turn of the century Riemann's theory of metrical continua, which had fallen so completely into oblivion, was revivified and deepened by Ricci and Levi-Civita; and that the work of these two decisively advanced the formulation of general relativity. However it seems to me that Levi-Civita's most important contribution lies in the following theoretical discovery: the most essential theoretical accomplishment of general relativity, namely the elimination of "rigid" space, i.e. of the inertial system, is only indirectly connected with the introduction of a Riemannian metric. The immediately essential conceptual element is the "displacement field" ($\Gamma^l_{ik}$) which expresses the infinitesimal displacement of vectors.

(23)

lassen sich nach allen Bildungsgesetze von Tensoren auf (27) in Verbindung mit Multiplikationen auffassen.

§11. Einige Spezialfälle von besonderer Bedeutung.

*Hilfssätze.*
Einige den Fundamentaltensor betreffende Differentialgesetze. Wir leiten zunächst einige im folgenden viel gebrauchte Hilfsgleichungen ab. Nach der Regel von der Differentiation der Determinanten ist

$$dg = g^{\mu\nu} g \, dg_{\mu\nu} = - g_{\mu\nu} g \, dg^{\mu\nu} \quad \cdots \cdots (28)$$

Die letzte Gleichsetzung form rechtfertigt sich durch die vorletzte, wenn man bedenkt, dass $g_{\mu\nu} g^{\mu\nu} = \delta_\mu^{\mu'}$, dass also $g_{\mu\nu} g^{\mu\nu} = 4$, folglich

$$g_{\mu\nu} \, dg^{\mu\nu} + g^{\mu\nu} \, dg_{\mu\nu} = 0.$$

Aus (28) folgt

$$\frac{1}{\sqrt{-g}} \frac{\partial \sqrt{-g}}{\partial x_\sigma} = \frac{1}{2} \frac{1}{g} \frac{\partial (-g)}{\partial x_\sigma} = \frac{1}{2} g^{\mu\nu} \frac{\partial g_{\mu\nu}}{\partial x_\sigma} = - \frac{1}{2} g_{\mu\nu} \frac{\partial g^{\mu\nu}}{\partial x_\sigma} \quad \cdots (29)$$

Aus

$$g_{\mu\sigma} g^{\nu\sigma} = \delta_\mu^\nu$$

folgt ferner durch Differentiation

$$g_{\mu\sigma} \frac{\partial g^{\nu\sigma}}{\partial x_\tau} = - g^{\nu\sigma} \frac{\partial g_{\mu\sigma}}{\partial x_\tau} \quad \cdots \cdots (30)$$

$$g_{\mu\sigma} \, dg^{\nu\sigma} = - g^{\nu\sigma} \, dg_{\mu\sigma}$$

bezw. $\quad g_{\mu\sigma} \frac{\partial g^{\nu\sigma}}{\partial x_\tau} = - g^{\nu\sigma} \frac{\partial g_{\mu\sigma}}{\partial x_\tau} \quad \Big\} (30)$

Durch gemischte Multiplikation mit $g^{\sigma\tau}$ bezw. $g_{\mu\lambda}$ erhält man hieraus (bei geänderter Bezeichnungsweise der Indizes)

$$d g^{\mu\nu} = - g^{\mu\alpha} g^{\nu\beta} \, dg_{\alpha\beta}$$
$$\frac{\partial g^{\mu\nu}}{\partial x_\sigma} = - g^{\mu\alpha} g^{\nu\beta} \frac{\partial g_{\alpha\beta}}{\partial x_\sigma} \quad \Big\} (31)$$

bezw.

$$d g_{\mu\nu} = - g_{\mu\alpha} g_{\nu\beta} \, dg^{\alpha\beta}$$
$$\frac{\partial g_{\mu\nu}}{\partial x_\sigma} = - g_{\mu\alpha} g_{\nu\beta} \frac{\partial g^{\alpha\beta}}{\partial x_\sigma} \quad \Big\} \cdots (32)$$

Die Beziehung (31) erlaubt eine Umformung, von der wir ebenfalls öfter Gebrauch zu machen haben. Gemäss ( ) ist

(23α)

$$\frac{\partial g_{\alpha\beta}}{\partial x_\sigma} = \left[ \begin{matrix} \alpha & \sigma \\ \beta \end{matrix} \right] + \left[ \begin{matrix} \beta & \sigma \\ \alpha \end{matrix} \right] \quad \cdots \cdots (33)$$

Setzt man dies in die zweite der Formeln 31 ein, so erhält man mit Rücksicht auf ( )

$$\frac{\partial g^{\mu\nu}}{\partial x_\sigma} = - \left( g^{\mu\tau} \left\{ \begin{matrix} \tau \sigma \\ \nu \end{matrix} \right\} + g^{\nu\tau} \left\{ \begin{matrix} \tau \sigma \\ \mu \end{matrix} \right\} \right) \quad \cdots (34)$$

durch Substitution der Ausdrücke von (34) in (29) erhält sich

$$\frac{1}{\sqrt{-g}} \frac{\partial \sqrt{-g}}{\partial x_\sigma} = \left\{ \begin{matrix} \sigma \\ \mu \end{matrix} \right\} \quad \cdots \cdots \cdots (29\alpha)$$

Einstein has already devoted a whole section (section 8) to the discussion of some properties of the "fundamental tensor," $g_{\mu\nu}$. He now lists a number of mathematical relations involving this tensor that will be used to introduce basic concepts of differential geometry.

### The *Entwurf* theory as an intermediate step toward the general theory of relativity

The mathematical concepts and methods presented in part B were already explored in the Zurich Notebook (p. 69 [16]). At the end of this effort in 1913, Einstein reached the conclusion that if the components of the metric tensor and only their first and second derivatives are included on the left-hand side of the gravitational field equation, then the requirement of energy-momentum conservation implies uniquely a system of equations that are not generally covariant. He then gave up the search for a generally covariant theory. Instead, he published, together with his mathematician friend Grossmann, the "Outline of a Generalized Theory of Relativity and of a Theory of Gravitation," which subsequently became known as the *Enwurf* (outline) theory. It was published in two parts: a "Physical Part" by Albert Einstein and a "Mathematical Part" by Marcel Grossmann. The field equations there are a direct outcome of the physical strategy (p. 64 [14]).

MPIWG Library

This theory was at the same time both a success and a failure. It was a success because Einstein and Grossmann had managed to derive a field equation for the new, complex representation of the gravitational potential, the metric tensor that was compatible with the Newtonian limit and thus seemed to stand on a firm physical basis. The *Entwurf* theory, however, was also a failure because it was not generally covariant, and it remained unclear to what extent it corresponded to Einstein's ambition to generalize the relativity principle to accelerated frames of reference. Einstein convinced himself at that time that this was the best that could be done. This left many questions open: Why, when general covariance was such a plausible heuristic requirement, was it impossible to implement it? Which were the reference systems preferred by his theory with Grossmann and why were they preferred? Between 1913 and 1915 Einstein attempted to answer these questions and to justify the limited covariance of the *Entwurf* theory.

In a letter to Lorentz in the spring of 1913, Einstein referred to this lack of general covariance as the "ugly dark spot" of the theory, but writing to Besso a year later, he expressed his complete satisfaction with the theory.

This is the longest (in size) page of the manuscript. Einstein finished page 23 and also page 24. Then he decided to add a few equations to page 23 and started a new page 23a. He realized that he did not need a whole page for this. He cut part of it and pasted it at the bottom of page 23. He then had to renumber the equations on page 24.

Divergenz des kontravarianten Vierervektors. Multipliziert man (26) mit dem kontravarianten Fundamentaltensor $g^{\mu\nu}$ (innere Multiplikation), so nimmt die rechte Seite nach Umformung des ersten Gliedes zunächst die Form an

$$\frac{\partial}{\partial x_\nu}(g^{\mu\nu}A_\mu) - A_\mu \frac{\partial g^{\mu\nu}}{\partial x_\nu} - \frac{1}{2}g^{\tau\alpha}\left(\frac{\partial g_{\mu\alpha}}{\partial x_\nu} + \frac{\partial g_{\nu\alpha}}{\partial x_\mu} - \frac{\partial g_{\mu\nu}}{\partial x_\alpha}\right)g^{\mu\nu}A_\tau.$$

Das letzte Glied dieses Ausdrucks kann gemäss (31) in die Form

$$\frac{1}{2}\frac{\partial g^{\tau\nu}}{\partial x_\nu}A_\tau + \frac{1}{2}\frac{\partial g^{\tau\mu}}{\partial x_\mu}A_\tau + \frac{1}{\sqrt{-g}}\frac{\partial\sqrt{-g}}{\partial x_\alpha}g^{\mu\nu}A_\tau$$

Da es auf die Benennung der Summationsindizes nicht ankommt heben sich die beiden ersten Glieder dieses Ausdruckes gegen das zweite des obigen weg; das letzte des er lässt sich mit dem ersten des obigen Ausdrucks vereinigen. Setzt man noch

$$g^{\mu\nu}A_\mu = A^\nu,$$

wobei $A^\nu$ ebenso wie $A_\mu$ ein frei wählbarer Vektor ist, so erhält man endlich

$$\Phi = \frac{1}{\sqrt{-g}}\frac{\partial}{\partial x_\nu}(\sqrt{-g}\,A^\nu) \quad \cdots\cdots\cdots (35)$$

Dieser Skalar ist die Divergenz des kontravarianten Vierervektors $A^\nu$.

"Rotation" des (kovarianten) Vierervektors. Das zweite Glied in (6) ist in den Indizes $\mu$ und $\nu$ symmetrisch. Es ist deshalb $A_\mu - A_{\nu\mu}$ ein besonders einfacher gebauter (antisymmetrischer) Tensor. Man erhält

$$B_{\mu\nu} = \frac{\partial A_\mu}{\partial x_\nu} - \frac{\partial A_\nu}{\partial x_\mu} \quad \cdots\cdots (36)$$

Antisymmetrische Erweiterung eines Sechservektors. Wendet man (27) auf einen antisymmetrischen Tensor zweiten Ranges $A_{\mu\nu}$ an, bildet hierzu die beiden durch zyklische Vertauschung der Indizes $\mu, \nu, \sigma$ entstehenden Gleichungen, und addiert diese drei Gleichungen, so erhält man den Tensor dritten Ranges

$$B_{\mu\nu\sigma} = A_{\mu\nu\sigma} + A_{\nu\sigma\mu} + A_{\sigma\mu\nu} = \frac{\partial A_{\mu\nu}}{\partial x_\sigma} + \frac{\partial A_{\nu\sigma}}{\partial x_\mu} + \frac{\partial A_{\sigma\mu}}{\partial x_\nu} \quad \cdots\cdots (37)$$

von welchem leicht zu beweisen ist, dass er antisymmetrisch ist.

Divergenz des Sechservektors. Multipliziert man (27) mit $g^{\mu\alpha}g^{\nu\beta}$ (gemischte Multiplikation), so erhält man ebenfalls einen Tensor. Das erste Glied der rechten Seite von (27) nimmt kann man in der Form

$$\frac{\partial}{\partial x_\sigma}(g^{\mu\alpha}g^{\nu\beta}A_{\mu\nu}) - g^{\mu\alpha}\frac{\partial g^{\nu\beta}}{\partial x_\sigma}A_{\mu\nu} - g^{\nu\beta}\frac{\partial g^{\mu\alpha}}{\partial x_\sigma}A_{\mu\nu}$$

schreiben. Ersetzt man $g^{\mu\alpha}g^{\nu\beta}A_{\mu\nu\sigma}$ durch $A_\sigma^{\alpha\beta}$, $g^{\mu\alpha}g^{\nu\beta}A_{\mu\nu}$ durch $A^{\alpha\beta}$ und ersetzt man in den umgeformten ersten Gliede $\frac{\partial g^{\nu\beta}}{\partial x_\sigma}$ und $\frac{\partial g^{\mu\alpha}}{\partial x_\sigma}$ vermittelst (34), so entsteht

**What is the *divergence* of a vector field? What are other vector field concepts?**

On this page, Einstein introduces the concept of the *divergence* of a vector. In classical physics, the divergence of a vector field at a point in space is the rate at which a physical entity "flows" out of a small volume surrounding that point. The divergence measures how much the vector field spreads out at each point and describes the strength of the source, in the case of an outgoing field, or the strength of the sink, in case of an incoming field. Since the source of a static electromagnetic field is the charge, just as the source of the static gravitational field is the mass, the divergence of each field is given by the charge or the mass enclosed within a small surface surrounding that point.

> The mathematical concept of divergence is related to the physical concept of conservation laws, because the divergence allows the behavior of a vector field as given by its sources to be related to the net flow through the surface. Let us demonstrate this using the example of the electric charge. In special relativity, the electric charge density and the electric current are the four components of a vector. The divergence of the charge-current vector in special relativity is the balance between the change in time of the charge density and the net flow of charge out of, or into, the region around a specific point. Unless charge is destroyed or created, this divergence vanishes, expressing the conservation law of electric charge. In general relativity, derivatives should in general be replaced by covariant derivatives (p. 78 [21]). It can be shown, however, that the divergence of a vector in general relativity has the same mathematical form as in special relativity.

In addition to the concept of divergence of a vector, Einstein introduces on this page three additional mathematical objects of tensor calculus:

- The *rotation* (*curl*) of a vector: a vector field, the lines of which surround an axis in space. Applying this operation to the electromagnetic potentials yields the antisymmetric electromagnetic field tensor (eq. 59, p. 110 [37]; see also p. 59 [11]).
- The antisymmetric extension of a six-vector: an antisymmetric tensor of rank 2 has six independent components (p. 59 [11]) and is sometimes called a six-vector. Applying this operation to the electromagnetic field tensor yields an antisymmetric tensor of rank 3 (eq. 60) representing Faraday's law and Gauss's law for magnetic fields (pp. 111 [37], 113 [38]).
- The divergence of a six-vector: the divergence of an antisymmetric contravariant tensor (a six-vector) is derived here as a step toward the derivation (on the next page) of the divergence of a mixed tensor of rank 2, which will appear in the energy-momentum conservation law.

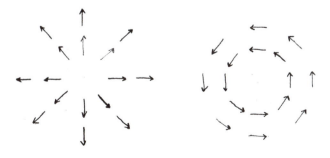

(25)

aus der rechten Seite von (27) ein sieben-gliedriger Ausdruck, von dem sich vier Glieder wegheben. Es bleibt übrig

$$A_\sigma^{\alpha\beta} = \frac{\partial A^{\alpha\beta}}{\partial x_\sigma} + \left\{ {\sigma\kappa \atop \alpha} \right\} A^{\kappa\beta} + \left\{ {\sigma\kappa \atop \beta} \right\} A^{\alpha\kappa} \dots \dots (38)$$

Es ist dies der Ausdruck für die Erweiterung eines kontravarianten Tensors zweiten Ranges, die sich entsprechend auch für kontravariante Tensoren höheren und niedrigeren Ranges bilden lässt.

Wir merken an, dass sich auf analogem Wege auch die Erweiterung eines gemischten Tensors $A_\mu^\alpha$ bilden lässt:

$$A_{\mu\sigma}^\alpha = \frac{\partial A_\mu^\alpha}{\partial x_\sigma} - \left\{ {\sigma\mu \atop \tau} \right\} A_\tau^\alpha + \left\{ {\sigma\tau \atop \alpha} \right\} A_\mu^\tau \dots \dots (39)$$

Durch Verjüngung von (38) bezüglich der Indizes $\beta$ und $\sigma$ (Multiplikation mit $\delta_\beta^\sigma$) erhält man den kontravarianten Vierervektor

$$A^\alpha = \frac{\partial A^{\alpha\beta}}{\partial x_\beta} + \left\{ {\beta\kappa \atop \beta} \right\} A^{\alpha\kappa} + \left\{ {\beta\kappa \atop \alpha} \right\} A^{\kappa\beta}$$

Wegen der Symmetrie von $\left\{ {\beta\kappa \atop \alpha} \right\}$ bezüglich der Indizes $\beta$ und $\kappa$ verschwindet das dritte Glied der rechten Seite, falls $A^{\alpha\beta}$ ein antisymmetrischer Tensor ist, was wir annehmen wollen. das zweite Glied lässt sich gemäss (29a) umformen. Man erhält also

$$A^\alpha = \frac{1}{\sqrt{-g}} \frac{\partial(\sqrt{-g}\, A^{\alpha\beta})}{\partial x_\beta} \dots \dots (40)$$

Dies ist der Ausdruck der Divergenz eines antisymmetrischen kontravarianten Tensors zweiten Ranges (Sechservektors).

Divergenz des gemischten Tensors zweiten Ranges. Bilden wir die Verjüngung von (39) bezüglich der Indizes $\alpha$ und $\sigma$, so erhalten wir mit Rücksicht auf (29a) den kovarianten

$$\sqrt{-g}\, A_\mu = \frac{\partial(\sqrt{-g}\, A_\mu^\sigma)}{\partial x_\sigma} - \left\{ {\sigma\mu \atop \tau} \right\} A_\sigma^\tau \dots \dots (41)$$

Führt man im letzten Gliede den kontravarianten Tensor $A^{\rho\sigma} = g^{\rho\tau} A_\tau^\sigma$ ein, so nimmt es die Form an

$$- \left[ {\sigma\mu \atop \rho} \right] \sqrt{-g}\, A^{\rho\sigma}$$

Ist ferner der Tensor $A^{\rho\sigma}$ ein symmetrischer, so reduziert sich dies

### What is the mathematical formulation of energy-momentum conservation in general relativity?

On this page, Einstein derives the divergence of a tensor of rank 2. To be more specific, he derives here the divergence of a mixed tensor. This is the form of the energy-momentum tensor that appears as a source (on the right-hand side) of the gravitational field equation.

> In special relativity, the mathematical procedure of deriving the divergence of a vector is now applied to every line of the matrix representing the tensor. Thus, the divergence of a tensor has four components: it is a vector. Let us demonstrate the meaning of this concept in the case of the energy-momentum tensor, mentioned briefly on page 12. The first three lines contain the density of the momentum components at a given point and the flow of these momentum components in the different spatial directions. The divergence of each line represents the balance between the change in time of the specific momentum component enclosed in a small volume around that point and the flow of that momentum component out of that volume. When there are no external forces acting on the system, the divergence vanishes, which represents the momentum conservation law. The divergence of the fourth line represents the balance between the change in time of the energy enclosed in that volume and its flow in the different directions. In a closed system, when no energy is supplied from an external source, this divergence vanishes, representing the energy conservation law.

The transition from special to general relativity ushers in a new element. The local changes (derivatives) of the components of the tensor with respect to the space and time coordinates have to be derived in a covariant way (covariant differentiation, p. 79). This procedure introduces new terms in addition to the ordinary combination of temporal and spatial derivatives. These terms involve the Christoffel symbols. Their physical significance (p. 75) relates the divergence of the electromagnetic tensor to the energy-momentum conservation in general relativity. We shall return to this point on page 99. Another, rather technical, point is the fact that we can assign to energy-momentum a covariant, a contravariant, or a mixed tensor. Any one of these forms can be changed into another one with the help of a metric tensor. Still, we have to make a choice and carefully keep track of it. The choice is made so that the equations are most comprehensible and the physical meaning of the quantities involved is most conveniently described. It turns out that this is best accomplished when the energy-momentum tensor is represented in the mixed form. This is why Einstein, on this page, talks about the "divergence of a mixed tensor of rank 2." In contrast with the case of special relativity, however, the vanishing of the covariant derivative of the energy momentum tensor cannot be interpreted as the true conservation law of a physical quantity.

(26)

auf $-\sqrt{-g}\,\frac{\partial g_{\varrho\sigma}}{\partial x_\mu}\,A^{\varrho\sigma}$. Hätte man statt $A^{\varrho\sigma}$ den ebenfalls symmetrischen kovarianten (Tensor $A_{\varrho\sigma} = g_{\varrho\alpha}\,g_{\sigma\beta}\,A^{\alpha\beta}$ eingeführt, so würde das letzte Glied vermöge (31) die Form $\sqrt{-g}\,\frac{\partial g^{\varrho\sigma}}{\partial x_\mu}\,A_{\varrho\sigma}$ annehmen. In dem betrachteten Symmetriefalle kann also (41) auch durch die beiden Formen

$$\sqrt{-g}\,A_\mu = \frac{\partial(\sqrt{-g}\,A_\mu^{\sigma})}{\partial x_\sigma} - \frac{\partial g_{\varrho\sigma}}{\partial x_\mu}\,\sqrt{-g}\,A^{\varrho\sigma} \quad \dots (41a)$$

und

$$\sqrt{-g}\,A_\mu = \frac{\partial(\sqrt{-g}\,A_\mu^{\sigma})}{\partial x_\sigma} + \frac{\partial g^{\varrho\sigma}}{\partial x_\mu}\,\sqrt{-g}\,A_{\varrho\sigma} \quad \dots (41b)$$

ersetzt werden, von denen wir im Folgenden Gebrauch zu machen haben.

### §12. Der Riemann-Christoffel'sche Tensor.

Wir fragen nun nach denjenigen Tensoren, welche aus dem Fundamentaltensor der $g_{\mu\nu}$ allein durch Differentiation gewonnen werden können. Die Antwort scheint zunächst auf der Hand zu liegen. Man setzt in (22) statt des beliebig gegebenen Tensors $A_{\mu\nu}$ den Fundamentaltensor der $g_{\mu\nu}$ ein und erhält dadurch einen neuen Tensor, nämlich die Erweiterung des Fundamentaltensors. Man überzeugt sich jedoch leicht, dass diese letztere identisch verschwindet. Man gelangt jedoch auf folgendem Wege zum Ziel. Man setze in (22)

$$A_{\mu\nu} = \frac{\partial A_\mu}{\partial x_\nu} - \begin{Bmatrix}\mu\nu\\ \varrho\end{Bmatrix}A_\varrho,$$

(bei etwas geänderter Benennung der Indizes)

d. h. die Erweiterung des Vierervektors $A_\nu$ ein. Dann erhält man den Tensor dritten Ranges

$$A_{\mu\sigma\tau} = \frac{\partial^2 A_\mu}{\partial x_\sigma \partial x_\tau}$$

$$-\begin{Bmatrix}\mu\sigma\\ \varrho\end{Bmatrix}\frac{\partial A_\varrho}{\partial x_\tau} - \begin{Bmatrix}\mu\tau\\ \varrho\end{Bmatrix}\frac{\partial A_\varrho}{\partial x_\sigma} - \begin{Bmatrix}\sigma\tau\\ \varrho\end{Bmatrix}\frac{\partial A_\mu}{\partial x_\varrho}$$

$$+\left[-\frac{\partial}{\partial x_\varrho}\begin{Bmatrix}\mu\sigma\\ \varrho\end{Bmatrix} + \begin{Bmatrix}\mu\tau\\ \alpha\end{Bmatrix}\begin{Bmatrix}\alpha\sigma\\ \varrho\end{Bmatrix} + \begin{Bmatrix}\sigma\tau\\ \alpha\end{Bmatrix}\begin{Bmatrix}\alpha\mu\\ \varrho\end{Bmatrix}\right]A_\varrho$$

Dieser Ausdruck ladet zur Bildung des Tensors $A_{\mu\sigma\tau} - A_{\mu\tau\sigma}$ ein. Denn dabei heben sich die folgende Terme das erste Glied des Ausdrucks für $A_{\mu\sigma\tau}$ gegen solche von $A_{\mu\tau\sigma}$ weg: das erste Glied, das vierte Glied, sowie das dem letzten Term in der eckigen Klammer entsprechende Glied. denn alle diese sind in $\sigma$ und $\tau$ symmetrisch. Gleiches gilt von der Summe des zweiten und dritten Gliedes. Wir erhalten also

## What is the geometric meaning of the Riemann-Christoffel tensor?

The Riemann-Christoffel tensor is an important mathematical object in differential geometry and in general relativity. It measures at each point to what extent the geometry in the neighborhood of that point is different from flat space (Euclidean space or Minkowski spacetime). It cannot be decided on the basis of the metric tensor alone. The metric tensor may also vary in flat space owing to the choice of the system of coordinates. The Riemann tensor, in contrast, is a convenient and straightforward diagnostic of the nature of space. In flat space it vanishes for any choice of the coordinate system. The Riemann tensor is better known today as the *Riemann curvature tensor*. Einstein does not mention the concept of curvature until October 1916, when for the first time he refers to the Riemann tensor as the Riemann tensor of curvature (p. 41 [A2]).

Curvature is a central concept in differential geometry. There are conceptually different ways to define it, associated with different mathematical objects, the metric tensor, and the affine connection. In our case, however, the affine connection may be derived from the metric. The "affine curvature" is associated with the notion of parallel transport of vectors as introduced by Levi-Civita. This is most simply illustrated in the case of a two-dimensional surface embedded in three-dimensional space. Let us take a closed curve on that surface and attach to a point on that curve a vector tangent to the surface. Let us now transport that vector along the curve, keeping it parallel to itself. When it comes back to its original position, it will coincide with the original vector if the surface is flat or deviate from it by a certain angle if the surface is curved. If one takes a small curve around a point on the surface, then the ratio of the angle between the original and the final vector and the area enclosed by the curve is the curvature at that point. The curvature at a point on a two-dimensional surface is a pure number.

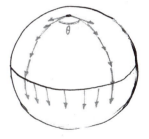

The notion of parallel transport applies also to the analysis of the curvature at a point in four-dimensional space, except that the situation is more complicated. The closed curve defining the track of the parallel transport can be located in one of infinitely many planes through that point. Two vectors are needed to specify the plane in which the parallel transport actually proceeds. Moreover, at the end of the loop, the angle between the final and the original vector will, in general, not be in the plane of the curve of the parallel transport. Thus, a second plane, defined by the initial and final vector, is necessary to specify the result of the process. The curvature is again the ratio between the angle of deviation and the area of the closed curve, but it now depends on the orientation of the two planes involved. Each of the four vectors defining these two planes contributes one index to the expression defining the curvature. This expression is the Riemann curvature tensor of rank 4.

Transporting a vector along a small closed loop necessitates keeping track of the changes of the vector as it is moved along the curve. Mathematically, this is equivalent to calculating derivatives of a vector. This has to be done by covariant differentiation (p. 79 [21]), which introduces Christoffel symbols. Moreover, the comparison of the changes in the vector at opposite sides of the closed curve also introduces changes in the Christoffel symbols themselves. Thus, the Riemann (curvature) tensor is a combination of Christoffel symbols and their derivatives (eq. 43, on the next page).

(27)

$$A_{\mu\sigma\tau} - A_{\mu\tau\sigma} = B^{\varsigma}_{\mu\sigma\tau} A_{\varsigma} \quad \cdots \cdots \cdots (42)$$

$$B^{\varsigma}_{\mu\sigma\tau} = -\frac{\partial}{\partial x_\tau}\left\{\begin{matrix}\mu\sigma\\\varrho\end{matrix}\right\} + \frac{\partial}{\partial x_\sigma}\left\{\begin{matrix}\mu\tau\\\varrho\end{matrix}\right\}$$

$$\left.\begin{matrix}\\ \\ \end{matrix}\right\} \cdots (43)$$

$$-\left\{\begin{matrix}\mu\sigma\\\alpha\end{matrix}\right\}\left\{\begin{matrix}\alpha\tau\\\varrho\end{matrix}\right\} + \left\{\begin{matrix}\mu\tau\\\alpha\end{matrix}\right\}\left\{\begin{matrix}\alpha\sigma\\\varrho\end{matrix}\right\}$$

Wesentlich ist an diesem Resultat, dass auf der rechten Seite von (42) nur die $A_{\varsigma}$ aber nicht mehrere Ableitungen auftreten. Aus dem Tensorcharakter von $A_{\mu\sigma\tau} - A_{\mu\tau\sigma}$ in Verbindung damit, dass $A_{\varsigma}$ ein frei wählbarer Vektor Vierervektor ist, folgt vermöge der Resultate des § 7, dass $B^{\varsigma}_{\mu\sigma\tau}$ ein Tensor ist (Riemann-Christoffel'scher Tensor).

Die mathematische Bedeutung dieses Tensors liegt in Folgendem. Wenn (das Kontinuum so beschaffen ist, dass es) vor dem Kontinuum ein Koordinatensystem so beschaffen Kontinuum gibt, bezüglich dessen die $g_{\mu\nu}$ Konstante sind, so verschwinden alle $R^{\varsigma}_{\mu\sigma\tau}$. Wählt man statt des ursprünglichen Koordinatensystems ein beliebiges neues, so werden die $g_{\mu\nu}$ in letzteren auf letzteres bezogenen $g_{\mu\nu}$ nicht Konstante sein. Der Tensorcharakter von $R^{\varsigma}_{\mu\sigma\tau}$ bringt es aber mit sich, dass diese Komponenten auch in dem beliebig gewählten Bezugssystem sämtlich verschwinden. Das Verschwinden des Riemann'schen Tensors ist also eine notwendige Bedingung dafür, dass durch geeignete Wahl des Bezugssystems die $g_{\mu\nu}$ konstant Konstanz der $g_{\mu\nu}$ herbeigeführt werden kann.[x] In unserem Problem entspricht dies dem Falle, dass bei passender Wahl des Koordinatensystems in endlichen Gebieten die spezielle Relativitätstheorie gilt.

Durch Verjüngung von (43) bezüglich der Indizes $\tau$ und $\varsigma$ erhält man den kovarianten Tensor zweiten Ranges

$$B_{\mu\nu} = R_{\mu\nu} + S_{\mu\nu}$$

$$B_{\mu\nu} = -\frac{\partial}{\partial x_\alpha}\left\{\begin{matrix}\mu\nu\\\alpha\end{matrix}\right\} + \left\{\begin{matrix}\mu\alpha\\\beta\end{matrix}\right\}\left\{\begin{matrix}\nu\beta\\\alpha\end{matrix}\right\}$$

$$\left.\begin{matrix}\\ \\ \\ \end{matrix}\right\} \cdots (44)$$

$$S_{\mu\nu} = \frac{\partial}{\partial x_\nu}\lg\frac{\partial^2\sqrt{-g}}{\partial x_\mu \partial x_\nu} - \left\{\begin{matrix}\mu\nu\\\alpha\end{matrix}\right\}\frac{\partial \lg\sqrt{-g}}{\partial x_\alpha}$$

Bemerkung über die Koordinatenwahl. Es ist schon in § 8 im Anschluss an Gleichung (18a) bemerkt worden, dass die Koordinatenwahl mit Vorteil so getroffen werden kann, dass $\sqrt{-g} = 1$ wird. Ein Blick auf die in den beiden letzten § Sgerlangten Gleichungen zeigt, dass durch eine solche Wahl

---

[x] Die Mathematiker haben bewiesen, dass diese Bedingung auch eine hinreichende ist.

## What was the "presumed gravitational tensor" and why was it abandoned?

The Riemann tensor appears in the Zurich Notebook in the search for covariant mathematical expressions constructed from derivatives of the metric tensor. There it is denoted by the four-index symbol (*ik,lm*). It is marked by the label "Grossmann tensor fourth rank," indicating the role of Grossman in bringing it to Einstein's attention.

Closely related to the Riemann tensor is the Ricci tensor, which is obtained from the Riemann tensor by contraction of the contravariant index with one of the covariant indices. The result is a sum of two covariant tensors of rank 2 (eq. 44). The advantage of choosing unimodular transformations (p. 71 [17]), implying $g = -1$, is now apparent. With this choice one of these terms vanishes. Therefore, with this choice of coordinates the formulation of the theory is greatly simplified. Einstein emphasizes that this choice of coordinates is adopted only for convenience and that after the theory is fully developed, it will be easy to revert to a generally covariant formulation. The Ricci tensor is a cornerstone of the general theory of relativity.

The Ricci tensor is used in the Zurich Notebook to generate candidates for the gravitational tensor in the field equations. Again, the name Grossman appears at the head of the page.

Einstein labels the second term of this tensor the "presumed gravitational tensor $T_{il}$." To accept this tensor as the gravitational tensor, he had to verify that it reduces to the Newtonian limit in the case of a weak static gravitational field, that it satisfies energy-momentum conservation, and that it allows for a generalization of the relativity principle. At the time of the Zurich Notebook, Einstein and Grossmann believed that this candidate failed the test and therefore dropped it. It was revived in November 1915 after the demise of the *Entwurf* theory.

(28)

die Bildungsgesetze der Tensoren eine bedeutende Vereinfachung erfahren. Besonders gilt dies für den soeben entwickelten Tensor
$B_{\mu\nu}$, welcher in der darzulegenden Theorie eine fundamentale Rolle
spielt. Die ins Auge gefasste Spezialisierung der Koordinatenwahl bringt
nämlich das Verschwinden von $S_{\mu\nu}$ mit sich, sodass sich der Tensor $B_{\mu\nu}$
auf $R_{\mu\nu}$ reduziert.

Ich will deshalb alle Beziehungen in der vereinfachten Form
angeben, welche die genannte Spezialisierung der Koordinatenwahl
mit sich bringt. Es ist dann ein Leichtes, auf die allgemein kovarianten
Gleichungen zurückzugreifen, falls dies in einem speziellen Falle
erwünscht erscheint.

### C. Theorie des Gravitationsfeldes.

§13. Bewegungsgleichung des materiellen Punktes im Gravitationsfeld. Ausdruck für die Feldkomponenten der Gravitation.

Ein frei beweglicher, äusseren Kräften nicht unterworfener
Körper bewegt sich nach der speziellen Relativitätstheorie geradlinig
und gleichförmig. Dies gilt auch nach der allgemeinen Relativitätstheorie
für einen Teil des vierdimensionalen Raumes, in welchem das Koordinatensystem $K_0$ so wählbar und so gewählt ist, dass die $g_{\mu\nu}$ die in (4) gegebenen
speziellen konstanten Werte haben.

Betrachten wir eben diese Bewegung von einem beliebig gewählten
Koordinatensystem $K_1$ aus, so bewegt er sich — von $K_1$ aus beurteilt —
nach den Überlegungen des §2 in einem Gravitationsfelde. Das Bewegungsgesetz mit Bezug auf $K_1$ ergibt sich leicht aus folgender Überlegung.
Mit Bezug auf $K_0$ ist das Bewegungsgesetz eine vierdimensionale Gerade,
also eine geodätische Linie. Da nun die geodätische Linie
unabhängig vom Bezugssystem definiert ist, wird ihre Gleichung
auch die Bewegungsgleichung des materiellen Punktes in bezug auf $K_1$ sein. Setzen wir

$$\Gamma_{\mu\nu}^{\tau} = -\{ {}^{\mu\;\nu}_{\;\tau} \}, \quad \ldots \ldots (45)$$

so lautet also die Gleichung der Punktbewegung inbezug auf $K_1$

$$\frac{d^2 x_\tau}{ds^2} = \Gamma_{\mu\nu}^{\tau} \frac{dx_\mu}{ds}\frac{dx_\nu}{ds}. \quad \ldots \ldots (46)$$

Wir machen nun die sehr naheliegende Annahme, dass dieses allgemein
kovariante Gleichungssystem die Bewegung des Punktes im Gravitationsfeld auch in dem Falle bestimmt, dass kein Bezugssystem $K_0$
existiert, bezüglich dessen in endlichen Räumen die spezielle

### When did Einstein lose faith in the *Entwurf* theory?

On November 4, 1915, Einstein announced that he had now found a way to realize his original vision of a general principle of relativity, which he saw embodied in the mathematical demand for general covariance. On that occasion, he wrote: "I lost trust in the field equations I had derived, and instead, looked for a way to limit the possibilities in a natural way. In this pursuit I arrived at the demand of general covariance, a demand from which I parted, though with a heavy heart, three years ago when I worked together with my friend Grossman. As a matter of fact we were then quite close to that solution of the problem, which will be given in the following." However, it took him until November 25th to finally obtain the solution.

The mathematical formalism required to derive the field equations of general relativity has now been fully described. Einstein continued to struggle with mathematics till the end of his life, not only in his efforts to unify gravity and electromagnetism into one theoretical framework but also to find an alternative to field theory, for instance, an algebraic theory for the description of reality.

In January 1943, Einstein received a letter from a young girl, Barbara Lee, from Washington. She confided in him: "I am below average in mathematics. I have to work longer in it than most of my friends. . . ." To this, Einstein responded: "Do not worry about your difficulties in mathematics; I can assure you that mine are still greater."

Part C is essentially a more detailed and comprehensive exposition of the theory presented to the Prussian Academy in four consecutive communications in November 1915, without explicit reference to that work and without any mention of his previous work with Grossman embodied in the Zurich Notebook and in the *Entwurf* theory.

### How does a particle move in a gravitational field?

The first step Einstein took was to explore the motion of a particle in a gravitational field. In classical physics a material particle moves, according to Newton's first law, along a straight line at constant velocity unless it is subjected to a force. Once Einstein recognized, as early as 1912, that gravitation is reflected in spacetime geometry, it became clear to him that the natural generalization of the straight line is the geodesic, following the straightest possible path. He therefore concluded that material particles move along geodetic lines unless they are subject to a force (other than gravitation).

The equation of motion of a particle in a gravitational field (eq. 46) is identical with the equation of the geodetic line (eq. 22, p. 76 [20]), except that Einstein has replaced the curly brackets notation, representing the Christoffel symbols, with the letter Γ. The left-hand side of the equation is the second derivative of the position of the particle with respect to the distance along the path of motion (the geodetic line). This distance is measured in units of time. Thus, the left-hand side is the acceleration of the particle. In relativity this time is called the *proper time*, and when the velocities are much smaller than the velocity of light, it reduces to ordinary time. According to Einstein, the Christoffel symbols represent the gravitational field, so when there is no gravitational field they vanish, the acceleration is zero, and particles move at constant velocity. In Newton's theory, acceleration depends on the gravitational field; eq. 46 replaces Newton's equation of motion in general relativity.

Einstein had already found the correct equation of motion in the summer of 1912. It was the field equation that posed the greatest challenge.

Relativitätstheorie gilt. Zu dieser Annahme sind wir umso berechtigter, als
(46) nur erste Ableitungen der $g_{\mu\nu}$ enthält, zwischen denen die Annahme
von der ~~Ex~~ auch im Spezialfalle der Existenz von $K_0$ keine Beziehungen
bestehen.*

Verschwinden die $T^\tau_{\mu\nu}$, so bewegt sich der Punkt gradlinig und gleichförmig; diese Grössen bedingen also die Abweichung der Bewegung
von der Gleichförmigkeit. Sie sind die Komponenten des Gravitationsfeldes.

§ 14. Die Feldgleichungen der Gravitation bei Abwesenheit von
Materie.

Wir unterscheiden im Folgenden zwischen "Gravitationsfeld"
und "Materie", in dem Sinne, dass alles ausser ~~als~~ "Materie" ~~bezeichnet~~ materiell
bezeichnet ~~wird~~, also nicht nur die "Materie" im üblichen Sinne
sondern auch das elektromagnetische Feld.

Unsere nächste Aufgabe ist es, die Feldgleichungen der Gravitation
bei Abwesenheit von Materie aufzusuchen. Dabei verwenden wir wieder
dieselbe Methode wie im vorigen § bei der Aufstellung der Bewegungsgleichung
des materiellen Punktes. Ein Spezialfall, in welchem die gesuchten Feld—
gleichungen jedenfalls erfüllt sein müssen, ist der der ursprünglichen
Relativitätstheorie, in dem die $g_{\mu\nu}$ gewisse konstante Werte haben. Dies
sei der Fall in einem gewissen endlichen Gebiete inbezug auf ein
bestimmtes Koordinatensystem $K_0$. Inbezug auf dieses System ~~sind~~
~~~~ verschwinden sämtliche Komponenten $B^\sigma_{\mu\sigma\tau}$ des Riemann'-
schen Tensors (Gleichung (43)). Diese verschwinden dann für das betrachtete Gebiet
auch bezüglich jedes anderen Koordinatensystems.

Die gesuchten Gleichungen des materiefreien Gravitationsfeldes
müssen also jedenfalls erfüllt sein, wenn alle $B^\sigma_{\mu\sigma\tau}$ verschwinden. Aber
diese Bedingung ist jedenfalls eine zu weitgehende. Denn es ist klar,
dass z. B. das von einem Massenpunkte in seiner Umgebung erzeugte
Gravitationsfeld ~~kann~~ sicherlich durch keine Wahl des Koordinatensystems
"wegtransformiert", d. h. auf den Fall konstanter $g_{\mu\nu}$ transformiert werden

Deshalb liegt es nahe, für das materiefreie Gravitationsfeld das
Verschwinden des aus dem Tensor $B^\sigma_{\mu\sigma\tau}$ abgeleiteten symmetrischen Tensors $B_{\mu\nu}$ zu verlangen.
Man erhält so 10 Gleichungen, ~~welche~~ für die 10 Grössen $g_{\mu\nu}$, welche
im speziellen erfüllt sind, wenn sämtliche $B^\sigma_{\mu\sigma\tau}$ verschwinden. ~~Man~~ Diese
~~erhält so mit~~ mit Rücksicht auf (44) bei der von uns ~~gewählt~~ getroffenen
Wahl für das Koordinatensystem für das materiefreie Feld die Gleichungen

$$\frac{\partial T^\alpha_{\mu\nu}}{\partial x_\alpha} + \Gamma^\alpha_{\mu\beta}\,T^\beta_{\nu\alpha} = 0 \left.\right\} \quad (47).$$
$$\sqrt{g} = 1.$$

* Erst zwischen den zweiten (und dritten) Ableitungen bestehen gemäss § 12 die Beziehungen
$B^\sigma_{\mu\sigma\tau} = 0.$

## What was Einstein's greatest challenge?

The greatest challenge of Einstein's search for a relativistic theory of gravitation was the search for a field equation that generalized Newton's theory in a plausible way and at the same time combined the insights following from the equivalence principle with those of special relativity. Here, Einstein's exposition reveals few traces of his earlier struggles and instead accentuates the mathematical elegance. We have already emphasized that gravitational fields are generated by matter. However, a gravitational field can exist without matter as its source. Einstein begins with this special case and shows that the mathematical framework of the absolute differential calculus almost immediately suggests a field equation. The actual field equation will then follow from the condition that it must represent a natural generalization of the case in which no matter is present.

Einstein's starting point is the Riemann-Christoffel tensor of rank 4, introduced earlier. Einstein had long realized that, in analogy with electromagnetism, the right-hand side of the equation, corresponding to the source of the field, must be the energy-momentum tensor of rank 2. Therefore, the left-hand side, which describes the geometry of spacetime representing the gravitational field, must be also a tensor of rank 2. We have already seen such a tensor, obtained by a contraction of the Riemann-Christoffel tensor (eq. 44, p. 90 [27]). It reduces to the Ricci tensor $R_{\mu\nu}$, confined to unimodular coordinates ($-g = 1$). This is the left hand-side of the field equation (47), where the Christoffel symbols are expressed by $\Gamma$. In the absence of matter, the right-hand side of this equation is zero.

At this point, however, Einstein renounced such physical arguments, suggesting that the field equation for the case of the absence of matter follows almost immediately from his "mathematical strategy." He first considered the field equation without matter, requiring it also to cover the case of special relativity or, more specifically, the case in which the components of the metric tensor are constant in a certain region and in a certain coordinate system. In this case, all components of the Riemann tensor vanish. Thus, the field equation has to be satisfied if the Riemann tensor vanishes. Einstein then argues that this condition would be too restrictive and that a natural way to relax it would be to require that only the components of the Ricci tensor vanish. This would give him the field equation for the matter-free case (eq. 47).

Einstein concludes the last two sections with a statement (on the next page), which is appropriate to present in his own words: "These equations (47), which proceed, by the method of pure mathematics, from the requirement of the general theory of relativity, give us, in combination with the equations of motion (46), to a first approximation Newton's law of attraction, and to a second approximation the explanation of the motion of the perihelion of the planet Mercury discovered by Leverrier. . . . These facts must in my opinion, be taken as convincing proof of the correctness of the theory."

Einstein is referring here to the calculation of the precession of the perihelion of planet Mercury, which was the content of his third communication presented in November 1915 to the Prussian Academy of Sciences. This remark seems to be a little out of context here, but Einstein sought to reassure the reader that he was on the right track.

(30)

Es muss darauf hingewiesen werden, dass *der Wahl dieser* ~~diesen~~ Gleichungen ein Minimum von Willkür anhaftet. Denn es gibt ausser $B_{\mu\nu}$ keinen Tensor zweiten Ranges, der aus den $g_{\mu\nu}$ und ~~deren~~ Ableitungen gebildet ist, keine höheren als zweite Ableitungen enthält, und in letzteren linear ist.[x]

Dass diese ~~ein~~ aus der Forderung der allgemeinen Relativität auf rein mathematischem Wege fliessenden Gleichungen in Verbindung mit den Bewegungsgleichungen (46) in erster Näherung das Newton'sche Attraktionsgesetz, in zweiter Näherung die Erklärung der von Leverrier entdeckten (nach Anbringung der Störungskorrektionen übrig bleibenden) Perihelbewegung des Merkur liefern, muss nach meiner Ansicht von der physikalischen Richtigkeit der Theorie überzeugen.

### §15. ~~Impuls-Energiesatz~~ Hamilton'sche Funktion für das Gravitationsfeld. Impuls-Energiesatz.

Um zu zeigen, dass die Feldgleichungen dem Impuls-Energiesatz entsprechen, ist es am bequemsten, sie in folgender Hamilton'scher Form zu schreiben

$$\delta\left\{\int \mathcal{H}\,d\tau\right\} = 0 \qquad \left.\begin{array}{l}\\[4pt]\end{array}\right\} \quad (47a)$$

$$\mathcal{H} = g^{\mu\nu}\,\Gamma^{\alpha}_{\mu\beta}\,\Gamma^{\beta}_{\nu\alpha}. \qquad \underset{\sqrt{-g}\,=\,1}{}$$

Dabei ~~sind~~ *verschwinden* die Variationen an den Grenzen des betrachteten begrenzten viersdimensionalen Integrationsraumes. Es ist zunächst zu zeigen, dass die Form (47a) den Gleichungen (47) äquivalent ist. Zu diesem Zweck betrachten wir $\mathcal{H}$ als Funktion der $g^{\mu\nu}$ und $g^{\mu\nu}_{\sigma}\left(=\dfrac{\partial g^{\mu\nu}}{\partial x_{\sigma}}\right)$. Dann ist zunächst

$$\delta\mathcal{H} = \Gamma^{\alpha}_{\mu\beta}\,\Gamma^{\beta}_{\nu\alpha}\,\delta g^{\mu\nu} + 2g^{\mu\nu}\Gamma^{\alpha}_{\mu\beta}\,\delta\Gamma^{\beta}_{\nu\alpha}$$

$$= -\Gamma^{\alpha}_{\mu\beta}\,\Gamma^{\beta}_{\nu\alpha}\,\delta g^{\mu\nu} + 2\Gamma^{\alpha}_{\mu\beta}\,\delta\left(g^{\mu\nu}\Gamma^{\beta}_{\nu\alpha}\right)$$

Nun ist aber

$$\delta\left(g^{\mu\nu}\Gamma^{\beta}_{\nu\alpha}\right) = -\frac{1}{2}\,\delta\left[g^{\mu\nu}g^{\beta\lambda}\left(\frac{\partial g_{\nu\lambda}}{\partial x_{\alpha}} + \frac{\partial g_{\alpha\lambda}}{\partial x_{\nu}} - \frac{\partial g_{\alpha\nu}}{\partial x_{\lambda}}\right)\right]$$

Die aus den beiden letzten Termen der runden Klammer hervorgehenden Terme ~~unterscheiden sich~~ (durch ihr ~~beziehen~~ *ausser* ~~und mir durch~~ sind von verschiedenem Vorzeichen und gehen aus einander ~~durch~~ (da die Benennung der Summationsindizes belanglos~~st~~ ist) durch Vertauschung der Indizes $\nu$ und $\beta$ hervor. Sie heben einander im Ausdruck für $\delta\mathcal{H}$ weg, weil sie mit der bezüglich der Indizes $\mu$ und $\beta$ symmetrischen Grösse $\Gamma^{\alpha}_{\mu\beta}$ multipliziert werden. Es bleibt also nur das erste Glied der runden Klammer zu berücksichtigen, sodass man mit Rücksicht auf (31) erhält

---

[x] Eigentlich lässt sich dies nur von dem Tensor $B_{\mu\nu} + \lambda\,g_{\mu\nu}\left(g^{\alpha\beta}B_{\alpha\beta}\right)$ behaupten, wobei $\lambda$ eine Konstante ist. Setzt man jedoch diesen gleich null, so kommt man wieder zu den Gleichungen $B_{\mu\nu} = 0$.

### What is the Lagrangian formalism, and what was its role in the genesis of general relativity?

After presenting his rapidly evolving new gravitation theory in November 1915, summarized in short, hastily written communications submitted to the Prussian Academy, Einstein exchanged several letters with friends in Leiden, the theoretical physicists Lorentz and Ehrenfest. They supported his work and general conclusions but had a number of queries, which Einstein tried to explain by referring to the November papers. Sometime around the end of January 1916 or later (we do not have the exact date), he realized that they deserved a detailed explanation of how he had derived the gravitational field equation. He wrote to Ehrenfest: "Today you should finally be content with me. I am delighted about the great interest you are devoting to this problem. I am not going to support myself at all on the papers but shall calculate everything for you." Einstein asked Ehrenfest to show this letter to Lorentz as well, and to return the letter to him, "because nowhere do I have these things so nicely in one place."

Einstein probably had this letter in front of him when he wrote part C of the manuscript. The derivations from section 15 onward closely follow this letter, apart from the Christoffel symbols, which are still denoted by curly bracketed expressions. It is interesting to note that it is in this letter that Einstein explicitly introduced the summation convention (p. 59).

It remained for him to show that the gravitational field alone, without a "matter" source defined by eq. (47), satisfies the energy-momentum conservation law. To this end, Einstein applied the Lagrangian formalism, which he had used in 1914 to derive the field equations of the *Entwurf* theory. He believed then that this derivation implied uniquely the *Entwurf* equations. It turned out that this conclusion was wrong, but at the time it solidified his confidence in the validity of the theory.

What is the Lagrangian formalism that played such an important role in the development of general relativity? Newton's mechanics is based on the concept of force, which is mathematically represented by a vector. Beginning with the work of Leibniz and its extension by Euler, Lagrange, and Hamilton, an alternative emerged that extends well beyond mechanics. This approach is based on the characterization of a physical process, such as the motion of a particle, as a quantity—usually called *Lagrangian* or *Hamiltonian*, but referred to here as *Hamiltonian*—and depends on the parameters describing the state of the system and their derivatives with respect to the space (or spacetime) coordinates. The dynamics is described by a variational principle instead of an equation of motion. According to the procedure, first introduced by Hamilton, the initial and final points of a motion are fixed, and the possible paths between them are characterized by a certain scalar quantity, referred to as the action, which may be obtained by the time integral of the Lagrangian. The actual motion of the particle (or the dynamics of the physical system) is given by an extremal (minimum or maximum) value of this integral. From Hamilton's "variational principle," it is then possible to derive certain differential equations for the motion, the so-called Euler-Lagrange equations. The Lagrangian formalism is one of the main tools for describing the dynamics of a vast variety of physical systems. It has been applied to derive the field equations of electromagnetism. In the case of fields, the Lagrangian looks like a scalar but is actually a scalar density, that is, a scalar multiplied by a factor depending on the coordinate transformation to ensure the preservation of volumes.

For Einstein, the Lagrangian formulation of the fundamental equations of general relativity had the advantage of allowing an elegant demonstration of their compatibility with the requirement of energy-momentum conservation, which had previously constituted a major problem in his search for the correct field equations.

$$\delta \mathcal{H} = - \Gamma_{\mu\beta}^{\alpha} \Gamma_{\nu\alpha}^{\beta} \, \delta g^{\mu\nu} \cdot - \Gamma_{\mu\beta}^{\alpha} \, \delta g_{\alpha}^{\mu\beta}.$$

Es ist also

$$\frac{\partial \mathcal{H}}{\partial g^{\mu\nu}} = - \Gamma_{\mu\beta}^{\alpha} \Gamma_{\nu\alpha}^{\beta}$$

$$\frac{\partial \mathcal{H}}{\partial g_{\sigma}^{\mu\nu}} = \Gamma_{\mu\nu}^{\sigma}.$$

$$\left.\right\} (48)$$

Die Ausführung der Variation in (47a) ergibt zunächst das Gleichungssystem

$$\frac{\partial}{\partial x_{\alpha}} \left( \frac{\partial \mathcal{H}}{\partial g_{\alpha}^{\mu\nu}} \right) - \frac{\partial \mathcal{H}}{\partial g^{\mu\nu}} = 0, \quad \cdots \sim (47b)$$

welches wegen (48) mit (47) übereinstimmt, was zu beweisen war. — Multipliziert man (47b) mit $g_{\sigma}^{\mu\nu} \xi$, so erhält man, ~~nach geläufigen und~~ ~~Umformung~~ weil

$$\frac{\partial g_{\sigma}}{\partial x_{\alpha}} = \frac{\partial g_{\alpha}^{\mu\nu}}{\partial x_{\sigma}}$$

und folglich

$$g_{\sigma}^{\mu\nu} \frac{\partial}{\partial x_{\alpha}} \left( \frac{\partial \mathcal{H}}{\partial g_{\alpha}^{\mu\nu}} \right) = \frac{\partial}{\partial x_{\alpha}} \left( g_{\sigma}^{\mu\nu} \frac{\partial \mathcal{H}}{\partial g_{\alpha}^{\mu\nu}} \right) - \frac{\partial \mathcal{H}}{\partial g_{\alpha}^{\mu\nu}} \frac{\partial g_{\alpha}^{\mu\nu}}{\partial x_{\sigma}}$$

die Gleichung

$$\frac{\partial}{\partial x_{\alpha}} \left( g_{\sigma}^{\mu\nu} \frac{\partial \mathcal{H}}{\partial g_{\alpha}^{\mu\nu}} \right) - \frac{\partial \mathcal{H}}{\partial x_{\sigma}} = 0$$

oder [*]

$$\frac{\partial t_{\sigma}^{\alpha}}{\partial x_{\alpha}} = 0$$

$$-2\kappa t_{\sigma}^{\alpha} = g_{\sigma}^{\mu\nu} \frac{\partial \mathcal{H}}{\partial g_{\alpha}^{\mu\nu}} - \delta_{\sigma}^{\alpha} \mathcal{H},$$

$$\left.\right\} (49)$$

oder wegen (48), ~~und~~ der zweiten Gleichung (47) und (34)

$$\kappa t_{\sigma}^{\alpha} = \frac{1}{2} \delta_{\sigma}^{\alpha} g^{\mu\nu} \Gamma_{\mu\beta}^{\alpha} \Gamma_{\nu\alpha}^{\beta} - g^{\mu\beta} \Gamma_{\mu\beta}^{\alpha} \Gamma_{\nu\sigma}^{\beta}$$

$$\kappa t_{\sigma}^{\alpha} = \frac{1}{2} \delta_{\sigma}^{\alpha} g^{\mu\nu} \Gamma_{\mu\beta}^{\alpha} \Gamma_{\nu\alpha}^{\beta} - g^{\mu\nu} \Gamma_{\mu\beta}^{\alpha} \Gamma_{\nu\sigma}^{\beta} \cdots (50)$$

Es ist ~~wohl~~ zu beachten, dass $t_{\sigma}^{\alpha}$ kein Tensor ist, dagegen gilt (49) für alle Koordinatensysteme, für welche $\sqrt{-g} = 1$ ist. Diese Gleichung drückt den Erhaltungssatz des Impulses und der Energie für das Gravitationsfeld aus. In der That liefert die Integration dieser Gleichung über ein dreidimensionales Volumen (die vier Gleichungen

[*] Der Grund der Einführung des Faktors $-2\kappa$ wird später angegeben.

## What happens to the energy-momentum conservation principle in the absence of matter, or can the gravitational field be a source of itself?

Einstein first applies the Lagrangian (he refers to it as the Hamiltonian) formalism to derive the gravitational field equation in the absence of matter (eq. 47). Next, he reformulates this equation, and in the process, a new set of quantities, $t_\mu^\nu$, appear that look like the components of a tensor. However, $t_\mu^\nu$ is not a tensor. Rather, it represents the energy and momentum of the gravitational field, which is not a covariant quantity but one that depends on the chosen reference frame. Still, the reformulation of the field equation with the help of this energy-momentum "complex" of the gravitational field played an important heuristic role for Einstein, suggesting the form in which the energy-momentum tensor of matter should be introduced into the field equation. The first of the equations (49) represents the conservation law of the energy-momentum of the gravitational field. In this form, it is valid in all systems of coordinates for which $g = -1$.

> Einstein was already aware in 1912 of the importance of the energy-momentum of the gravitational field and its role in the gravitational field equation while he was working on the theory of the static gravitational field. His first version of the theory violated energy-momentum conservation. When he added a term to correct this, he realized that it represented the energy-momentum of the gravitational field itself. This insight shaped his further search for the field equation.

## Who was Einstein's main competitor?

> Einstein's final phase in completing his theory of general relativity in November 1915 was a solitary effort. He had little correspondence on the subject, apart from that with the mathematician David Hilbert with whom he communicated on the progress of their respective work. Hilbert had a long-standing interest in fundamental issues within his program of an axiomatization of physics. He was drawn to the electrodynamic theory of matter published by Gustav Mie in 1912. Hilbert's hope was that particles such as the electron could be derived from the electromagnetic field. They would be represented by singular point-like structures of the electromagnetic field lines.
>
> After Einstein's visit in Göttingen, at Hilbert's invitation, in the summer of 1915, Hilbert attempted to integrate Mie's theory of matter with Einstein's theory of gravitation, but still taking the *Entwurf* theory as his starting point. Hilbert and Einstein exchanged criticism and preliminary results, directly and possibly indirectly through others. Still, it is clear that Einstein completed the last steps of his theory along the pathways of his previous research.
>
> In November 1915, Hilbert was close to completing his integrated theory of electrodynamics and relativity and became Einstein's competitor for priority in formulating the field equation of the gravitational field.

(32)

$$\frac{d}{dx_4}\left\{\int t_\sigma^4\, dV\right\} = \int \left(t_\sigma^1 \alpha_1 + t_\sigma^2 \alpha_2 + t_\sigma^3 \alpha_3\right) dS, \quad \cdots (49\alpha)$$

wobei $\alpha_1, \alpha_2, \alpha_3$ die Richtungskosinus der nach innen gerichteten Normale eines Flächenelementes (von der Grösse $dS$) der Begrenzung (im Sinne der euklidischen Geometrie) bedeuten. Man erkennt hierin den Ausdruck des "Erhaltungssatzes in üblicher Fassung. Die Grössen $t_\sigma^\alpha$ bezeichnen wir als die "Energie-Komponenten" des Gravitationsfeldes. Ich will nun die Gleichungen (42) noch in einer dritten Form angeben, die einer lebendigen Erfassung unseres Gegenstandes besonders dienlich ist. Durch Multiplikation der Feldgleichungen (42) ergeben sich in der mit $g^{\nu\sigma}$ dieser "gemischten" Form. Beachtet man, dass

$$g^{\nu\sigma}\frac{\partial \Gamma_{\mu\nu}^{\alpha}}{\partial x_\alpha} = \frac{\partial}{\partial x_\alpha}\left(g^{\nu\sigma}\Gamma_{\mu\nu}^{\alpha}\right) - \frac{\partial g^{\nu\sigma}}{\partial x_\alpha}\Gamma_{\mu\nu}^{\alpha},$$

welche Grösse wegen (34) gleich

$$\frac{\partial}{\partial x_\alpha}\left(g^{\nu\sigma}\Gamma_{\mu\nu}^{\alpha}\right) \mp - g^{\nu\beta}\Gamma_{\sigma\beta}^{\alpha}\Gamma_{\mu\nu}^{\alpha} - g^{\sigma\beta}\Gamma_{\beta\alpha}^{\nu}\Gamma_{\mu\nu}^{\alpha}$$

oder (nach $\mp$ geänderter Benennung der Summationsindizes) gleich

$$\frac{\partial}{\partial x_\alpha}\left(g^{\sigma\beta}\Gamma_{\mu\beta}^{\alpha}\right)\left(- g^{\nu\sigma}\Gamma_{\mu\beta}^{\alpha}\Gamma_{\nu\alpha}^{\beta} - g^{\nu\sigma}\Gamma_{\mu\beta}^{\sigma}\Gamma_{\nu\mu}^{\beta}\right)$$

Das dritte Glied dieses Ausdrucks hebt sich weg gegen das aus dem zweiten Glied der Feldgleichungen (42) entstehende; sodass anstelle des zweiten Gliedes dieses Ausdrucks lässt sich nach (50)

$$\kappa\left(t_\mu^\sigma - \tfrac{1}{2}\delta_\mu^\sigma\, t\right)$$

setzen $(t = t_\alpha^\alpha)$. Man erhält also anstelle der Gleichungen (42)

$$\left.\frac{\partial}{\partial x_\alpha}\left(g^{\sigma\beta}\Gamma_{\mu\beta}^{\alpha}\right) = -\kappa\left(t_\mu^\sigma - \tfrac{1}{2}\delta_\mu^\sigma\, t\right)\right\} \quad \cdots (51)$$
$$\sqrt{-g} = 1$$

§16. Allgemeine Fassung der Feldgleichungen der Gravitation.

Die im vorigen § aufgestellten Feldgleichungen sind der für materiefreie Räume Feldgleichung

$$\Delta \varphi = 0$$

der Newton'schen Theorie zu vergleichen. Wir haben die Gleichungen aufzusuchen, welche der Poisson'schen Gleichung

$$\Delta \varphi = 4\pi K \rho$$

entspricht, wobei $\rho$ die Dichte der Materie bedeutet. Die spezielle Relativitätstheorie hat zu dem

### How can the field equation without matter be generalized to include matter?

Einstein has used the variational method to derive the field equation in the absence of matter (eq. 47). The problem now is to generalize this equation in the presence of matter. To this end, Einstein proceeds to cast it in yet a third form, "which is particularly suitable for a lively apprehension of our subject." In this form (eq. 51) the expression identified as the energy-momentum of the gravitational field appears on the right-hand side of the equation and acts as the source of the field. All that is necessary now is to add to the right-hand side the energy-momentum of matter and give it the same form in which the energy-momentum expression of the gravitational field enters the equation.

As the last step, before generalizing this equation to include ordinary matter, Einstein looks back at the gravitational field equation in Newtonian theory (the so-called Poisson equation) in which the mass density $\rho$ is the source of the field $\varphi$.

In his *Autobiographical Notes*, Einstein comments on the role of the Poisson equation in the emergence of the field concept in physics. This equation is a way of expressing Newton's famous law of gravitation in terms of a space-filling potential that gives everywhere rise to a field, and hence to a force, behaving according to this law. But while the force law itself, describing how the gravitational force changes with distance, seems arbitrary, the Poisson equation relates the gravitational potential to a property of space itself, thus anticipating the later concept of "field," as Einstein points out in a discussion of Newtonian mechanics and its concept of force:

> The law of motion is precise, although empty as long as the expression for the forces is not given. For postulating the latter, however, there is an enormous degree of arbitrariness, especially if one drops the requirement, which is not very natural in any case, that the forces depend only on the coordinates (and not, for example, on their derivatives with respect to time). Within the framework of that theory alone, it is entirely arbitrary that the forces of gravitation (and electricity), which come from one point, are governed by the potential function $(1/r)$. Additional remark: it has long been known that this function is the spherically symmetrical solution of the simplest (rotation-invariant) differential equation $\Delta\Phi = 0$; it would therefore not have been far-fetched to regard this as a clue that this function was to be considered as resulting from a spatial law, an approach that would have eliminated the arbitrariness in the force law. This is really the first insight that suggests a turning away from the theory of action at a distance, a development that—prepared by Faraday, Maxwell, and Hertz—really begins only later in response to the external pressure of experimental data.

(33)

Ergebnis geführt, dass die träge Masse nichts anderes ist als Energie, welche ihren vollständigen mathematischen Ausdruck in einem symmetrischen Tensor, dem Energie-tensor findet. Wir werden daher auch in der allgemeinen Relativitätstheorie einen Energietensor (der Materie) $T_\sigma^\alpha$ einzuführen haben, der wie die $t_\sigma^\alpha$ der Gleichungen (49) und (50) des Gravitations-feldes gemischten Charakter haben wird, aber zu einem symmetrischen kovarianten Tensor gehören wird.

Wie dieser Energietensor (entsprechend der Dichte $\varrho$ in der Poisson'schen Gleichung) in die Feldgleichungen der Gravitation einzuführen ist, lehrt das Gleichungssystem (51). Betrachtet man nämlich ein vollständiges System (z. B. das Sonnensystem) so wird die Gesamtmasse des Systems, also auch seine gesamte gravitierende Wirkung von der Gesamtenergie des Systems, also von der ponderabeln und Gravitationsenergie zusammen abhängen. Dies wird sich dadurch ausdrücken lassen, dass man in (51) anstelle der Energiekomponenten des Gravitations-feldes allein die Summen $t_\mu^\sigma + T_\mu^\sigma$ der Energiekomponenten von Materie und Gravitationsfeld einführt. Man erhält so statt (51) die Tensorgleichung

$$\frac{\partial}{\partial x_\alpha}\left(g^{\sigma\beta}\Gamma_{\mu\beta}^\alpha\right) = -\kappa\left[\left(t_\mu^\sigma + T_\mu^\sigma\right) - \tfrac{1}{2}\delta_\mu^\sigma(t + T)\right] \quad \Big\} (52)$$

$$\sqrt{-g} = 1,$$

wobei $T = T_\mu^\mu$ gesetzt ist (Laue'scher Skalar). Anstatt dieser sind die gesuchten allgemeinen Feldgleichungen der Gravitation in gemischter Form. Anstelle von (47) ergibt sich daraus rückwärts das System

$$\frac{\partial \Gamma_{\mu\nu}^\alpha}{\partial x_\alpha} + \Gamma_{\mu\beta}^\alpha \Gamma_{\nu\alpha}^\beta = -\kappa\left(T_{\mu\nu} - \tfrac{1}{2}g_{\mu\nu}T\right) \quad \Big\} (53).$$

$$\sqrt{-g} = 1.$$

Es muss zugegeben werden, dass diese Einführung des Energietensors der Materie durch das Relativitätspostulat allein nicht gerechtfertigt wird, deshalb haben wir sie im Vorigen aus der Forderung abgeleitet, dass die Energie des Gravitationsfeldes in gleicher Weise gravitierend wirken soll, wie jegliche Energie anderer Art. Der stärkste Grund für die Wahl der vorstehenden Gleichungen liegt aber darin,

×  $g_{\alpha\tau}T_\sigma^\alpha = T_{\sigma\tau}$ und $g^{\sigma\beta}T_\sigma^\alpha = T^{\alpha\beta}$ sollen symmetrische Tensoren sein.

## The gravitational field equation—at last!

In general relativity the mass density, appearing on the right-hand side of the Poisson equation, is replaced by the energy-momentum tensor of matter. On the previous page, Einstein has prepared the way in which this tensor has to be introduced as a source term of the field equation. He then requires that the energy-momentum of matter and the energy-momentum of the field enter the field equation on the same footing. This requirement is the main motivation for assuming the particular form (eq. 52, which is then easily transformed to eq. 53) of Einstein's field equation. He then further clarifies that the main justification for this postulated field equation is the physical consequences deduced from it. Specifically, this will lead to the conservation of the total energy-momentum of matter and of the gravitational field (on the next page).

Equation (53) represents the triumphal achievement of Einstein's effort in searching for a generally covariant equation for the gravitational field. He recalled this achievement as a result of the mathematical strategy, rather than as the outcome of an intricate search for the convergence of physical and mathematical strategies. The left-hand side of the field equation is the explicit form of the Ricci tensor, which Einstein had already identified as the central element in general relativity in 1912. The source of the field on the right-hand side is introduced in a different way than before, namely, with an additional term: the trace of the energy-momentum tensor (the sum of its diagonal components).

If one insists on the usual form of the right-hand side, one has to modify the left-hand side of the equation, adding the trace of the Ricci tensor. The modified expression on the left-hand side is known as the Einstein tensor. The latter is the standard form of the gravitational field equation as known today. This form was adopted by Einstein by the year 1918.

$$R_{\mu\nu} - \tfrac{1}{2}\, g_{\mu\nu}\, R = -k T_{\mu\nu}$$

Years later, in 1936, Einstein described this equation as follows:

"The theory . . . is similar to a building, one wing of which is made of fine marble (the left wing of the equation), but the other wing of which is built of low-grade wood (the right wing of the equation). The phenomenological representation of matter is, in fact, only a crude substitute for a representation which would do justice to all known properties of matter."

dass sie zur Folge haben, dass für die Komponenten der Totalenergie Erhaltungsgleichungen (des Impulses und der Energie) gelten, welche den Gleichungen (49) und (49a) genau entsprechen. Dies soll im Folgenden dargethan werden.

§ 17. Die Erhaltungssätze im allgemeinen Falle.

Wir bilden an Gleichung (52) zunächst die Verjüngung nach den

Die Gleichung (52) ist leicht so umzuformen, dass auf der rechten Seite das zweite Glied wegfällt. Man verjüngt (52) nach den Indizes $\mu$ und $\sigma$ und subtrahiere die so erhaltene, mit $\frac{1}{2} \delta_\mu^\sigma$ multiplizierte Gleichung von (52). Es ergibt sich

$$\frac{\partial}{\partial x_\lambda}\left(g^{\sigma\beta} T_{\mu\,\beta}^{\,\alpha} - \frac{1}{2}\delta_\mu^\sigma g^{\lambda\beta} T_{\lambda\,\beta}^{\,\alpha}\right) = -\kappa\left(t_\mu^{\,\sigma} + T_\mu^{\,\sigma}\right) \cdots (52a)$$

An dieser Gleichung bilden wir die Operation $\frac{\partial}{\partial x_\sigma}$. Es ist

$$\frac{\partial^2}{\partial x_\lambda \partial x_\sigma}\left(g^{\sigma\beta} T_{\mu\,\beta}^{\,\kappa}\right) = -\frac{1}{2}\frac{\partial^2}{\partial x_\lambda \partial x_\sigma}\left[g^{\sigma\beta} g^{\kappa\lambda}\left(\frac{\partial g_{\mu\lambda}}{\partial x_\beta} + \frac{\partial g_{\beta\lambda}}{\partial x_\mu} - \frac{\partial g_{\mu\beta}}{\partial x_\lambda}\right)\right]$$

Das erste und das dritte Glied der runden Klammer liefern Beiträge, die einander wegheben, wie man erkennt, wenn man im Betrage des dritten Gliedes die Summationsindices $\alpha$ und $\sigma$ einerseits, $\beta$ und $\lambda$ andererseits vertauscht. Das zweite Glied lässt sich nach (31) umformen, sodass man erhält

$$\frac{\partial^2}{\partial x_\lambda \partial x_\sigma}\left(g^{\sigma\beta} T_{\mu\,\beta}^{\,\alpha}\right) = \frac{1}{2}\frac{\partial^3 g^{\alpha\beta}}{\partial x_\lambda \partial x_\beta \partial x_\mu} \quad \cdots (54)$$

Das zweite Glied der linken Seite von (52a) liefert zunächst

$$-\frac{1}{2}\frac{\partial^2}{\partial x_\lambda \partial x_\mu}\left(g^{\lambda\beta} T_{\lambda\,\beta}^{\,\alpha}\right)$$

oder

$$\frac{1}{4}\frac{\partial^2}{\partial x_\lambda \partial x_\mu}\left[g^{\lambda\beta} g^{\alpha\delta}\left(\frac{\partial g_{\delta\lambda}}{\partial x_\beta} + \frac{\partial g_{\delta\beta}}{\partial x_\lambda} - \frac{\partial g_{\lambda\beta}}{\partial x_\delta}\right)\right]$$

Das vom letzten Glied der runden Klammer herrührende Glied verschwindet wegen (29) bei der von uns getroffenen Koordinatenwahl. Die beiden anderen lassen sich zusammenfassen und liefern wegen (31) zusammen

$$-\frac{1}{2}\frac{\partial^3 g^{\alpha\beta}}{\partial x_\alpha \partial x_\beta \partial x_\mu},$$

sodass mit Rücksicht auf (54) die Identität

$$\frac{\partial^2}{\partial x_\lambda \partial x_\sigma}\left(g^{\sigma\beta} T_{\mu\,\beta}^{\,\alpha} - \frac{1}{2}\delta_\mu^\sigma g^{\lambda\beta} T_{\lambda\,\beta}^{\,\alpha}\right) \equiv 0 \quad \cdots (55)$$

besteht. Man erhält deshalb

$$\frac{\partial\left(t_\mu^{\,\sigma} + T_{\mu}^{\,\sigma}\right)}{\partial x_\sigma} = 0 \quad \cdots 56.$$

### How is the conservation principle satisfied in a way that Einstein did not expect in the early stages of development of his theory?

Einstein now shows that the postulated field equation satisfies energy-momentum conservation, stressing that it is the sum of energy-momentum of matter and of the gravitation field that is conserved and not the two components separately (eq. 56). This is the resolution of the problem that disturbed Einstein during the long search for a generally covariant field equation. Originally, the energy-momentum conservation principle was a separate requirement that seemed to be incompatible with general covariance. To satisfy the conservation principle, he had to restrict the allowed coordinate systems and thereby give up general covariance.

Einstein's four communications to the Prussian Academy of Sciences in November 1915 began with his return to exactly the equation he had considered three years earlier with Grossmann but abandoned because he had not been able to prove its compatibility with energy-momentum conservation. Now, however, on the basis of the variational formalism, Einstein was able to solve this problem. But one condition remained, resulting from the requirement of energy-momentum conservation. At this stage, the condition $-g = 1$ still served as a coordinate restriction (p. 71). Thus, in this "November field equation" there was still a discrepancy between covariance and conservation laws. In his next paper, published a week later, Einstein attempted to use this discrepancy to argue in favor of an electromagnetic origin of matter. Assuming a purely electromagnetic nature of matter imposed a condition on its energy-momentum tensor that would solve this puzzle. It was on the basis of this modified theory that Einstein calculated the Mercury perihelion shift, finding the correct value. At the same time, he discovered in the course of his calculation that he had to revise his ideas about how the new theory would yield the limiting case of Newtonian gravitation theory. This discovery then finally opened the gate for introducing the energy-momentum of matter in a slightly different way into the field equation, with the implication that conservation laws no longer imposed an additional condition that restricted general covariance.

Thus, Einstein's attempts to solve the discrepancy between covariance and conservation laws paved his way to the final field equation, which he published on November 25th. From a generally covariant field equation, energy-momentum conservation can be derived as a consequence. Einstein concluded the last paper stating, with relief, that "the postulate of general relativity cannot reveal to us anything new and different about the essence of various processes in nature than what the special theory of relativity already taught us. The opinions I recently voiced here in this regard were in error."

On the day that Einstein made the first presentation of his theory to the Prussian Academy of Sciences, he wrote a letter to his son Hans Albert: "You can learn a lot of good things from me that no one else can offer you. The things I have gained from so much strenuous work should be of value not only to strangers but especially to my own boys. In the last few days I completed one of the finest papers of my life. When you are older I will tell you about it. . . . I am often so engrossed in my work that I forget to eat lunch."

Aus unsern Feldgleichungen der Gravitation geht also hervor, dass den Erhaltungssätzen des Impulses und der Energie Genüge geleistet ist. Man sieht dies am einfachsten nach der Betrachtung die zu Gleichung (49a) führt, nur hat man hier anstelle der Energiekomponenten $t_\sigma^5$ des Gravitationsfeldes die Gesamt-Energiekomponenten von Materie und Gravitationsfeld einzuführen.

§18. Der Impuls- Energiesatz für die Materie als Folge der Feldgleichungen.

Multipliziert man (53) mit $\frac{\partial g^{\mu v}}{\partial x_\sigma}$, so erhält man auf dem in §15 eingeschlagenen Wege mit Rücksicht auf das Verschwinden von $g_{\mu v}\frac{\partial g^{\mu v}}{\partial x_\sigma}$ die Gleichung

$$\frac{\partial t_\sigma^\alpha}{\partial x_\alpha} = \frac{1}{2}\frac{\partial g^{\mu v}}{\partial x_\sigma} T'_{\mu v},$$

oder mit Rücksicht auf (50)

$$\frac{\partial T_\sigma^\alpha}{\partial x_\alpha} + \frac{1}{2}\frac{\partial g^{\mu v}}{\partial x_\sigma} T'_{\mu v} = \sigma \ldots\ldots(52)$$

bei der getroffenen Wahl für das Koordinatensystem

Ein Vergleich mit (41 b) zeigt, dass diese Gleichung nichts anderes aussagt als das Verschwinden der Divergenz des Tensors der Energiekomponenten der Materie. Physikalisch zeigt das Auftreten des zweiten Gliedes der linken Seite, dass für die Materie allein Erhaltungssätze des Impulses und der Energie im eigentlichen Sinne nicht gelten, bezw. nur dann gelten, wenn die $g^{\mu v}$ konstant sind, d. h. wenn die Feldstärken der Gravitation verschwinden. Dies zweite Glied ist ein Ausdruck für Impuls bezw. Energie, welche pro Volumen und Zeiteinheit vom Gravitationsfelde auf die Materie übertragen werden. Dies tritt noch klarer hervor, wenn man statt (52) im Sinne von (41) schreibt

$$\frac{\partial T_\sigma^\alpha}{\partial x_\alpha} = -\Gamma_{\sigma\beta}^\alpha T_\alpha^\beta \ldots\ldots(52a).$$

energetische)

Die rechte Seite drückt die Einwirkung des Gravitationsfeldes auf die Materie aus.

Die Feldgleichungen der Gravitation enthalten also gleichzeitig vier Bedingungen, welchen der materielle Vorgang zu genügen hat. Sie liefern die Gleichungen des materiellen Vorganges vollständig, wenn letzterer durch vier voneinander unabhängige

## Do physical conservation laws follow from symmetries in nature?

In the last step of the deductive construction of his theory, Einstein established a bridge to the work of Hilbert, incorporating one of its central mathematical results—the relation between conservation and covariance that was later generalized in Noether's theorem—into his newly established theory of gravitation. In the context of the *Entwurf* theory, Einstein had developed this relation in his own terms. He later elaborated on it in a paper he published in October 1916 to be discussed in detail in the annotations to the appendix (pp. 130–139 [A1–A5]). In the manuscript, he acknowledged Hilbert's publication with a footnote (at the bottom of the next manuscript page).

> Emmy Noether was a German mathematician working alongside Hilbert in Göttingen, one of the great centers of mathematics in those days. She is known for her groundbreaking contributions to abstract algebra and theoretical physics. In physics, she is best known for what is now called Noether's theorem. The theorem says that every symmetry is associated with a corresponding conservation law. Noether's theorem has been considered one of the most important mathematical theorems guiding the development of modern physics. In the context of general relativity, general covariance can be interpreted as a basic symmetry characterizing the spacetime geometry of the universe, and energy-momentum conservation is the associated conservation law implied by Noether's theorem. After her death in 1935, Einstein wrote about her in a letter to the *New York Times*: "In the judgment of the most competent living mathematicians, Fräulein Noether was the most significant creative mathematical genius thus far produced since the higher education of women began."

In conclusion of part C of the manuscript, it is interesting to look back at the difficulties that Einstein faced in 1912–13, and how these were resolved in his "November theory." In his search for a gravitational field equation, the requirement of being able to choose a coordinate system in which the Newtonian limit (as he understood it) could be realized and the requirement of energy-momentum conservation became entangled. Einstein finally succeeded in disentangling these problems when he realized that a field equation based on the November tensor could be made compatible with energy-momentum conservation by imposing just one weak coordinate restriction. This allowed him to see that what he needed to recover the Poisson equation from his almost generally covariant theory was just a coordinate condition and no longer a coordinate restriction. He managed to decouple the problem of energy-momentum conservation from the problem of recovering the Poisson equation, thus untying the knot that had hindered progress toward a generally covariant theory of gravitation in the Zurich Notebook.

> On the day following Einstein's presentation of the final version of his theory, he wrote to his friend Zangger: "The theory is of unequalled beauty. But just one of my colleagues has really understood it, and that one is cleverly trying to appropriate it." David Hilbert did not appropriate it, although he did walk in Einstein's footsteps. In his paper, he acknowledged that Einstein deserved the credit for discovering the general theory of relativity: "The differential equations of gravitation that result here are, as it seems to me, in agreement with the magnificent theory of general relativity established by Einstein."

(36)

Differentialgleichungen charakterisierbar ist.[x]

### D. Die „materiellen" Vorgänge.

Die unter B entwickelten mathematischen Hilfsmittel setzen uns ohne Weiteres in den Stand, die physikalischen Gesetze der Materie (Hydrodynamik, Maxwell'sche Elektrodynamik) so zu verallgemeinern, dass sie in die allgemeine Relativitäts-theorie hineinpassen. Dabei ergibt das allgemeine Relativitäts-prinzip zwar keine weitere Einschränkung der Möglichkeiten, aber es lehrt den Einfluss des Gravitationsfeldes auf alle Prozesse exakt kennen, ohne dass irgendwelche neue Hypothese eingeführt werden müsste.

Diese Sachlage bringt es mit sich, dass über die physikalische Natur der Materie (im engeren Sinne) nicht notwendig bestimmte Voraussetzungen eingeführt werden müssen. Insbesondere kann die Frage offen bleiben, ob die Theorie des elektromagnetischen Feldes und des Gravitationsfeldes zusammen eine hinreichende Basis für die Theorie der Materie liefern oder nicht. Das allgemeine Relativitäts-postulat kann uns hierüber im Prinzip nichts lehren. Es muss sich bei dem Ausbau der Theorie zeigen, ob Elektromagnetik und Gravitationslehre zusammen leisten können, was erstere allein nicht gelingen will.

### §19. Euler'sche Gleichungen für reibungslose adiabatische Flüssigkeiten.

Es seien $p$ und $\varrho$ zwei Skalare, von denen wir ersteren als „Druck", letzteren als die „Dichte" einer Flüssigkeit bezeichnen; zwischen ihnen bestehe eine Gleichung. Der kontravariante symmetrische Tensor

$$T^{\alpha\beta} = -g^{\alpha\beta} p + \varrho \frac{dx_\alpha}{ds} \frac{dx_\beta}{ds} \quad \ldots \ldots (58)$$

sei der kontravariante Energietensor der Flüssigkeit. Zu ihm gehört der kovariante Tensor

$$T_{\mu\nu} = -g_{\mu\nu} p + g_{\mu\alpha} \frac{dx_\alpha}{ds} g_{\nu\beta} \frac{dx_\beta}{ds} \varrho \quad \ldots \ldots (58a)$$

sowie der gemischte Tensor[xx]

$$T_\sigma^\alpha = -\delta_\sigma^\alpha p + g_{\sigma\beta} \frac{dx_\beta}{ds} \frac{dx_\alpha}{ds} \varrho \quad \ldots \ldots (58b)$$

Setzt man die rechten Seite von (58b) in (57a) ein, so erhält man die Euler'schen hydrodynamischen Gleichungen der allgemeinen

---

[x] Vgl. hierüber D. Hilbert. Nachr. d. K. Gesellsch. d. W. z. Göttingen. Math. phys. Klasse. 1915. S. 3

[xx] Für einen mitbewegten Beobachter, der im unendlich Kleinen ein Bezugsystem im Sinne der speziellen Relativitätstheorie benutzt, ist die Energiedichte $T_4^4$ gleich $\varrho - p$. Hierin liegt die Definition von $\varrho$. Es ist also $\varrho$ nicht konstant für eine inkompressible Flüssigkeit.

## How do established theories in physics, like hydrodynamics and electromagnetism, fit into the new theory of gravitation?

In part C, Einstein had derived the gravitational field equation (eq. 53) in which the covariant energy-momentum tensor $T_{\mu\nu}$ represents all the physical entities, except for the gravitational field itself. He refers to everything outside of the gravitational field as "matter." With the help of the mathematical tools developed in part B, he now examines how two such "material" examples (hydrodynamics and electromagnetism) fit into the framework of general relativity. He emphasizes that without the need for any new physical hypothesis, the effect of the gravitational field on these material phenomena can be determined. The reformulation of the well-known special relativistic equations of hydrodynamics and electrodynamics in this new context does not lead to any further conditions on the material processes described by them.

Einstein leaves open whether the combination of the new theory of gravitation with electrodynamics will lead to a new theory of matter, which contemporary scientists such as Gustav Mie had tried to build on electrodynamics alone. Einstein alludes here to the challenge of bringing electromagnetism and gravitation into the framework of a single theory, without mentioning explicitly Hilbert's effort in that direction. This precedes by several years Einstein's own serious effort to embark on the search for such a unified theory. Though unsuccessful, this effort occupied him until the end of his life.

In November 1914, Einstein published a review article on "The Formal Foundation of the General Theory of Relativity," summarizing the Einstein-Grossmann *Entwurf* theory. In this article, he derived the equations of hydrodynamics and the field equations of the electrodynamics of moving bodies, referring to both cases as "the laws of material processes." These equations remain unchanged also in the final theory. The only difference is that the field potentials $g_{\mu\nu}$, which appear in these equations, have to be derived from the correct gravitational field equations. Part D of the present manuscript is an abbreviated version of the corresponding sections in the 1914 article, except that the treatment of the electromagnetic energy-momentum tensor is greatly simplified. Specifically, Einstein got rid of the "six-vectors," which complicated previous formulations of this subject (also on the next page).

In addition to the mass and energy density, the inner pressure $p$ is also a source of the gravitational field and appears in the energy-momentum tensor. Inner pressure is related to the random motion of particles constituting the massive medium. This motion carries energy and, like every type of energy, contributes to the gravitational field. The generally relativistic Euler equations are the four equations obtained by applying equation (57a) to the mixed form of this tensor (eq. 58b).

Around the mid-eighteenth century, Leonhard Euler published a set of equations governing the motion of incompressible fluids. These equations essentially expressed the conservation of mass and momentum in any hydrodynamic process. An energy conservation equation was obtained almost a century later, but all three equations today are referred to as Euler's equations. With the discovery of special relativity, mass, energy, and momentum are combined into one energy-momentum tensor. Euler's equations in special relativity are obtained by setting the four components of the covariant divergence of this tensor to zero.

(37)

Relativitätstheorie. Diese lösen das Bewegungsproblem im Prinzip vollständig; denn die vier Gleichungen (57 a) zusammen mit der gegebenen Gleichung zwischen $p$ und $\varsigma$ und der Gleichung

$$g_{\alpha\beta}\frac{dx_\alpha}{ds}\frac{dx_\beta}{ds} = 1$$

genügen ~~zusammen~~ bei gegebenen $g_{\alpha\beta}$ zur Bestimmung der 6 Unbekannten

$$p, \varsigma, \frac{dx_1}{ds}, \frac{dx_2}{ds}, \frac{dx_3}{ds}, \frac{dx_4}{ds}.$$

Sind auch die $g_{\mu\nu}$ unbekannt, so kommen hiezu noch die Gleichungen (53). Das sind 11 Gleichungen zur Bestimmung der 10 Funktionen $g_{\mu\nu}$, sodass diese überbestimmt scheinen. Es ist indessen zu beachten, dass die Gleichungen (57 a) in den Gleichungen (53) bereits enthalten sind, sodass ~~unsere~~ letztere nur mehr 7 unabhängige Gleichungen repräsentieren. Diese Unbestimmtheit hat ihren guten Grund darin, dass die weitgehende ~~Freiheit~~ in der Wahl der Koordinaten es mit sich bringt, dass das Problem mathematisch in ~~weitgehendem~~ solchem ~~Masse~~ Grade unbestimmt bleibt, dass drei der Raumfunktionen beliebig gewählt werden können.[x]

§20. Maxwell'sche Elektromagnetische ~~total~~ Gleichungen für das Vakuum.

Es seien $\varphi_\nu$ die Komponenten eines kovarianten Viervektors, des ~~Viervektorverdes~~ elektromagnetischen Potentials. Aus ihnen bilden wir ~~die~~ gemäss (36) Komponenten $F_{\rho\sigma}$ des kovarianten Sechservektors des elektromagnetischen Feldes gemäss dem Gleichungssystem

$$F_{\rho\sigma} = \frac{\partial \varphi_\rho}{\partial x_\sigma} - \frac{\partial \varphi_\sigma}{\partial x_\rho}. \quad \cdots \quad (59)$$

Aus (59) folgt, dass das Gleichungssystem

$$\frac{\partial F_{\rho\sigma}}{\partial x_\tau} + \frac{\partial F_{\sigma\tau}}{\partial x_\rho} + \frac{\partial F_{\tau\rho}}{\partial x_\sigma} = 0 \quad \cdots \quad (60)$$

erfüllt ist, ~~welches gemäss~~ dessen linke Seite gemäss (37) ein antisymmetrischer Tensor dritten Ranges ist. Das System (60) enthält also im Wesentlichen 4 Gleichungen die ausgeschrieben wie folgt lauten

$$\frac{\partial F_{23}}{\partial x_4} + \frac{\partial F_{34}}{\partial x_2} + \frac{\partial F_{42}}{\partial x_3} = 0$$
$$\frac{\partial F_{34}}{\partial x_1} + \frac{\partial F_{41}}{\partial x_3} + \frac{\partial F_{13}}{\partial x_4} = 0$$
$$\frac{\partial F_{41}}{\partial x_2} + \frac{\partial F_{12}}{\partial x_4} + \frac{\partial F_{24}}{\partial x_1} = 0$$
$$\frac{\partial F_{12}}{\partial x_3} + \frac{\partial F_{23}}{\partial x_1} + \frac{\partial F_{31}}{\partial x_2} = 0.$$

$\left.\right\} (60a)$

[x] Bei Verzicht auf die Koordinatenwahl gemäss $g = -1$ bleiben ~~vier~~ Raumfunktionen frei wählbar, entsprechend den vier willkürlichen ~~Funktionen~~ über die man bei der Koordinatenwahl frei verfügen kann.

$z.\,45/3$

**How did Maxwell represent the laws of electromagnetism by mathematical equations and how are these equations affected by gravitation?**

In this section, Einstein essentially reproduces his "New Formal Interpretation of Maxwell's Field Equations of Electrodynamics," which he published in the reports of the Prussian Academy a few months earlier. That publication constitutes a simplification of the exposition of the same subject in his 1914 review article "The Formal Foundation of General Theory of Relativity." It allows an immediate transition to the general theory of relativity. Einstein was pleased by this new simplified covariant formulation of Maxwell's equations and corresponded about it with Lorentz. In September 1915, he was happy to inform Lorentz that "I have also found a proof for the validity of the energy-momentum conservation principle for the electromagnetic field taking gravitation into consideration, as well as a simple covariant theoretical representation of the vacuum equations in which the "dual" six-vector concept proves unessential."

> By the mid-nineteenth century, the British physicist Michael Faraday and the Scottish physicist and mathematician James Clerk Maxwell had introduced the concept of electric and magnetic fields. At that time, several physical laws about the relations between electric charges and currents, and electric and magnetic fields, had been empirically established:
>
> - Gauss's law, describing the static electric field, produced by (positive or negative) electric charges.
> - Gauss's law for magnetic fields, stating that there are no magnetic charges (so-called magnetic monopoles) in nature. They always appear in pairs (dipoles) like the "south pole" and "north pole" of a magnetic compass needle. Static magnetic fields, produced by magnetic materials, are generated by such dipoles.
> - Ampère's law, stating that an electric current produces a magnetic field. The magnetic field lines surround the current in a closed loop.
> - Faraday's law, stating that a time-varying magnetic field generates an electric field. The electric field lines surround the magnetic field in a closed loop.
>
> Maxwell added to these laws an additional statement that a magnetic field can be produced also by a time-varying electric field. Thus, a time-varying electric field acts like an electric current (the so-called *displacement current*). This was not required by any experimental result but was a consequence of Maxwell's ingenious insight. Around 1862, he published the first version of a set of mathematical equations describing all these laws. These equations are the celebrated Maxwell equations. They constitute the greatest unification scheme in the history of classical physics, combining electricity, magnetism, and optics into one framework. They predict the existence of electromagnetic waves propagating through space with the velocity of light, $c$.

In line with Minkowski's formulation of special relativity, it became possible to derive Maxwell's equations from the antisymmetric electromagnetic tensor $F_{\mu\nu}$, constructed from the derivatives of the electromagnetic potentials (electric and magnetic fields). Einstein begins this process on the present page. In 1912 Friedrich Kottler was the first to write down Maxwell's equations in generalized coordinates.

## 38

Dies Gleichungssystem entspricht dem zweiten Gleichungssystem Maxwells. Man erkennt dies sofort, indem man setzt

$$\begin{aligned}\mathcal{F}_{23} &= f_x & \mathcal{F}_{14} &= n_x \\ \mathcal{F}_{31} &= f_y & \mathcal{F}_{24} &= n_y \\ \mathcal{F}_{12} &= f_z & \mathcal{F}_{34} &= n_z\end{aligned}\quad\Bigg\}\ (61)$$

Dann kann man statt (60a) in üblicher Schreibweise der drei-dimensionalen Vektoranalysis setzen

$$\begin{aligned}\frac{\partial f}{\partial t} + \mathrm{rot}\, n &= 0 \\ \mathrm{div}\, f &= 0\end{aligned}\quad\Bigg\}\ (60b)$$

Das erste Maxwell'sche System erhalten wir durch Verallgemeinerung der Minkowski angegebenen Form. Wir führen den zu $\mathcal{F}_{\alpha\beta}$ gehörigen kontravarianten Sechservektor

$$\mathfrak{F}^{\mu\nu} = g^{\mu\alpha} g^{\nu\beta} \mathcal{F}_{\alpha\beta} \ \cdots (62)$$

ein sowie den kontravarianten Vierervektor $\mathfrak{J}^\mu$ der elektrischen Vakuum — Stromdichte, dann mag man mit Rücksicht auf (40) gegenüber beliebigen Substitutionen von der Determinante 1 (gemäss der von uns getroffenen Koordinatenwahl) invariante Gleichungssystem ansetzen

$$\frac{\partial \mathfrak{F}^{\mu\nu}}{\partial x_\nu} = \mathfrak{J}^\mu \ \cdots\cdots (63)$$

Setzt man nämlich

$$\begin{aligned}\mathfrak{F}^{23} &= f_x' & \mathfrak{F}^{14} &= -n_x' \\ \mathfrak{F}^{31} &= f_y' & \mathfrak{F}^{24} &= -n_y' \\ \mathfrak{F}^{12} &= f_z' & \mathfrak{F}^{34} &= -n_z',\end{aligned}\quad\Bigg\}\ (64)$$

welche Grössen im Fall der speziellen Relativitäts-theorie den Grössen $f_x \cdots n_z$ gleich sind, und ausserdem

$$\mathfrak{J}^1 = i_x,\ \mathfrak{J}^2 = i_y,\ \mathfrak{J}^3 = i_z,\ \mathfrak{J}^4 = \varrho$$

so erhält man anstelle von (63)

$$\begin{aligned}\mathrm{rot}\, f' - \frac{\partial n'}{\partial t} &= i \\ \mathrm{div}\, n' &= \varrho\end{aligned}\quad\Bigg\}\ \cdots (63a)$$

Die Gleichungen (60), (62) und (63) bilden also die Verallgemeinerung der Maxwell'schen Feldgleichungen des Vakuums bei der von uns bezüglich der Koordinatenwahl getroffenen Festsetzung.

## What was the role of "ether" in prerelativity physics and why did Einstein eventually think that space without ether is unthinkable?

The velocity of light $c$ appears explicitly in Maxwell's equations. This raised the question: with respect to what is this velocity measured? Physicists assumed that a weightless invisible medium permeates all space and provides an absolute reference frame with respect to which the velocity of light is defined. This medium was also meant to transmit the forces that act on electric charges and magnetic poles. Einstein's basic postulate underlying his special theory of relativity is the constancy of the speed of light in all reference frames, which move with constant velocity with respect to each other (inertial reference frames). This postulate determines the laws of transformation between inertial reference frames. Maxwell's equations are invariant under such transformations.

In the special theory of relativity, the velocity of light is the same in all inertial frames of reference, and the concept of ether becomes superfluous. Thus, Einstein removed it completely from his theoretical framework. However, in 1920 he distanced himself from this view to the point of identifying the dynamic spacetime of general relativity with ether, though not as an absolute reference frame. In a lecture delivered in October 1920 in Leiden on "Ether and the Theory of Relativity," he stated: "We may say that according to the general theory of relativity space is endowed with physical qualities; in this sense, therefore, there exists an ether. According to the general theory of relativity space without ether is unthinkable; . . . but this ether may not be thought of as endowed with the quality characteristic of ponderable media, as consisting of parts which may be tracked through time. The idea of motion may not be applied to it." The last sentence in this quotation relates to the fact that the ether concept that may be used to interpret the properties of spacetime in general relativity cannot be understood as a mechanical medium, unlike the notion of ether in prerelativity physics that Einstein discarded.

The distinction between covariant and contravariant vectors and tensors applies also in special relativity. Maxwell's equations describing Gauss's law for magnetic fields and Faraday's law (see the preceding page) are derived from the covariant form of the electromagnetic tensor $F_{\mu\nu}$, while the other two equations are derived from its contravariant form. Einstein keeps track of this difference by marking the electric and magnetic field components in the equations derived from the contravariant form of the tensor $F$, with a prime. Because of the simple form of the metric tensor of special relativity, the primed and nonprimed values of the magnetic and electric fields are the same, and this distinction is usually ignored in treatments of special relativity.

Einstein denotes the electric field by the letter $n$ and the magnetic field by $f$. In the printed version, the more conventional notation at that time, $e$ and $h$, is used.

In his *Autobiographical Notes*, written at the age of 67, Einstein expresses his fascination with Maxwell's theory: "The most fascinating subject at the time that I was a student was Maxwell's theory. What made this theory appear revolutionary was the transition from action at a distance to fields as the fundamental variable."

## 39

Die Energiekomponenten ~~des~~ elektromagnetischen Feldes. Wir bilden das innere Produkt

$$K_\sigma = F_{\sigma\mu} J^\mu \quad \ldots \ldots (65)$$

Seine Komponenten lauten gemäss (61) in dreidimensionaler Schreibweise

$$\left. \begin{array}{l} K_1 = \varrho \, n_x + [i, f]_x \\ - - - - - - - \\ - - - - - - \\ K_4 = -(i, n). \end{array} \right\} \quad (65a)$$

Es ist $K_\sigma$ ein (kovarianter) (Vierervektor, dessen Komponenten gleich sind dem negativen Impuls bezw. der Energie, welche pro Zeit- und Volumeneinheit auf das elektromagnetische Feld ~~von den elektrischen Massen~~ (übertragen werden.) Sind die elektrischen Massen frei, d. h. unter dem alleinigen Einfluss des elektromagnetischen Feldes, so wird der kovariante Vierervektor $K_\sigma$ verschwinden.

Um die (Energie-) Komponenten $T_\sigma^\nu$ des elektromagnetischen Feldes zu erhalten, brauchen wir nur der Gleichung $K_\sigma = 0$ die Gestalt der Gleichung (57) zu geben. Aus (63) und (65) ergibt sich zunächst

$$K_\sigma = F_{\sigma\mu} \frac{\partial F^{\mu\nu}}{\partial x_\nu} = \frac{\partial}{\partial x_\nu}\left( F_{\sigma\mu} F^{\mu\nu} \right) - F^{\mu\nu} \frac{\partial F_{\sigma\mu}}{\partial x_\nu}.$$

Das zweite Glied der rechten Seite gestattet vermöge (60) die Umformung

$$F^{\mu\nu} \frac{\partial F_{\sigma\mu}}{\partial x_\nu} = -\frac{1}{2} F^{\mu\nu} \frac{\partial F_{\mu\nu}}{\partial x_\sigma} = -\frac{1}{2} g^{\mu\alpha} g^{\nu\beta} F_{\alpha\beta} \frac{\partial F_{\mu\nu}}{\partial x_\sigma},$$

welch letzterer Ausdruck aus Symmetriegründen auch in der Form

$$-\frac{1}{4}\left[ g^{\mu\alpha} g^{\nu\beta} F_{\alpha\beta} \frac{\partial F_{\mu\nu}}{\partial x_\sigma} + g^{\mu\alpha} g^{\nu\beta} \frac{\partial F_{\alpha\beta}}{\partial x_\sigma} F_{\mu\nu} \right]$$

geschrieben werden kann. Dafür aber lässt sich setzen

$$-\frac{1}{4} \frac{\partial}{\partial x_\sigma}\left( g^{\mu\alpha} g^{\nu\beta} F_{\alpha\beta} F_{\mu\nu} \right) + \frac{1}{4} F_{\alpha\beta} F_{\mu\nu} \frac{\partial}{\partial x_\sigma}\left( g^{\mu\alpha} g^{\nu\beta} \right).$$

Das erste dieser Glieder lautet in kürzerer Schreibweise

$$-\frac{1}{4} \frac{\partial}{\partial x_\sigma}\left( F^{\mu\nu} F_{\mu\nu} \right),$$

## What was von Laue's crucial role?

Einstein is now ready to construct the energy-momentum tensor of electromagnetism. He begins with the four equations that specify the force acting on the system and the energy supplied to it. With the help of mathematical tools developed in part B, these equations can be transformed to a form in which the components of the energy-momentum tensor can be identified: eqs. (66) and (66a) on the next page. In a closed system, the right-hand side of eq. (66) is equal to zero, and one can show that it is identical with the equations expressing energy-momentum conservation, derived in part C (eq. 57).

We have already mentioned that part D of the manuscript is based on Einstein's treatment of the "the laws of material phenomena" (hydrodynamics and electromagnetism). There he mentions that his treatment of the subject uses the formulation of Minkowski-Laue, namely, the Minkowski tensorial formulation of special relativity and the von Laue formulation of flow of continuous matter in special relativity.

Max von Laue made important contributions to the development of relativity theory, in particular to the implications of Einstein's mass-energy relation $E = mc^2$ for a deeper understanding of relativistic continuum dynamics. According to the mass-energy relation, stresses, as familiar from elasticity theory or hydrodynamics, also embody energy and thus may change the inertial properties of a moving body. In fact, all forms of energy must have mass. In contrast, in classical physics stresses have no effect on the motion of a body, for example, when they are produced by a pair of equal and opposite forces acting along the same line. Even when dealing with an extended body, in classical physics it is always possible to describe the effect of forces on its overall motion—leaving aside deformations—by a single quantity, its inertial mass. In relativity, however, in general there is no longer a single quantity such as mass characterizing the inertial behavior of an extended physical system. The work of Max von Laue around 1911 made it clear that for this purpose no fewer than 10 functions are required, which together form the components of a geometric object in spacetime called the "stress-energy tensor" or the "energy-momentum tensor." As early as 1912 Einstein realized that this tensor must play the crucial role of the source-term in his field equation for the gravitational field, taking the place of the mass density in its classical counterpart, the Poisson equation for Newton's gravitational potential.

Laue himself, however, at first remained skeptical with regard to Einstein's theory. In August 1913 he wrote (to Schlick): "The extraordinary, in fact inconceivable complexity of the theory confirms my rejection. Fortunately, one of its most direct consequences, the bending of light rays near the sun, can be checked already in 1914 during the solar eclipse. Then the theory will likely pass away peacefully."

das zweite ergibt nach Ausführung der Differentiation nach einiger Umformung

$$-\tfrac{1}{2}\, \mathfrak{F}^{\mu\tau}\mathfrak{F}_{\mu\nu}\, g^{\nu\rho}\frac{\partial g_{\sigma\tau}}{\partial x_\sigma}$$

Nimmt man alle drei berechneten Glieder zusammen, so erhält man die Relation

$$K_\sigma = \frac{\partial \mathfrak{T}_\sigma^{\ \nu}}{\partial x_\nu} - \tfrac{1}{2}\, g^{\tau\mu}\frac{\partial g_{\mu\nu}}{\partial x_\sigma}\mathfrak{T}_\tau^{\ \nu}, \quad \ldots\ldots (66)$$

wobei

$$\mathfrak{T}_\sigma^{\ \nu} = -\mathfrak{F}_{\sigma\alpha}\mathfrak{F}^{\nu\kappa} + \tfrac{1}{4}\delta_\sigma^\nu \mathfrak{F}_{\alpha\beta}\mathfrak{F}^{\alpha\beta} \quad \ldots\ldots (66a)$$

Die Gleichung (66) ist für verschwindendes $K_\sigma$ wegen (30) mit (52) bezw. (52a) gleichwertig. Es ist also $\mathfrak{T}_\sigma^{\ \nu}$ der sind also die Kompo $\mathfrak{T}_\sigma^{\ \nu}$ die Energiekomponenten des elektromagnetischen Feldes. Mit Hilfe von (61) und (64) zeigt man leicht, dass diese Energiekomponenten des elektromagnetischen Feldes im Falle der speziellen Relativitätstheorie die wohlbekannten Maxwell-Pointing'schen Ausdrücke ergeben.

§21.)

**E. Newtons Theorie als erste Näherung.**

§21. Gesichtspunkte für die Aufstellung

Wie schon mehrfach erwähnt, ist die spezielle Relativitätstheorie als Spezialfall der allgemeinen dadurch charakterisiert, dass die $g_{\mu\nu}$ die konstanten Werte (4) haben. Dies bedeutet nach dem Vorherigen eine völlige Vernachlässigung der Gravitationswirkungen. Eine der Wirklichkeit näher liegende Approximation erhalten wir, indem wir den Fall betrachten, dass die $g_{\mu\nu}$ von den Werten (4) nur um (gegen 1) kleine Grössen abweichen, wobei wir unendlich kleine Grössen zweiten und höheren Grades vernachlässigen (erster Genäuerungspunkt der Approximation).

Ferner soll angenommen werden, dass in dem betrachteten Gebiete die $g_{\mu\nu}$ im Unendlichen bei passender Wahl der Koordinaten den Werten (4) zustreben, d. h. wir betrachten Gravitationsfelder, welche als ausschliesslich durch im Endlichen befindliche Materie erzeugt werden können.

Man könnte annehmen, dass diese Vernachlässigungen aus auf Newtons Theorie bringen müssten. Indessen bedarf es hiefür noch der approximativen Behandlung der Grundgleichungen nach einem zweiten Gesichtspunkte. Wir fassen die Bewegung eines Massenpunktes gemäss den Gleichungen (46) ins Auge. Im Falle der speziellen Relativitätstheorie können die Komponenten beliebige Werte annehmen, welche der Bedingung das bedeutet, dass

$$\left(\frac{dx_1}{ds}\right)^2 + \left(\frac{dx_2}{ds}\right)^2 + \left(\frac{dx_3}{ds}\right)^2 < 1$$

## How can the validity of the theory be tested experimentally?

In parts A, B, C, and D of this manuscript, Einstein has completed the theoretical framework and the mathematical apparatus of his new relativistic theory of gravitation: the general theory of gravity. Each of these parts has a title; part E does not. Einstein immediately refers to section 21, having deleted the previous title and replaced it with a simpler, shorter one. Part E discusses the first tests of validity of the general theory of relativity: explanation of the observed precession of the perihelion of the planet Mercury, the bending of light rays from distant stars in the gravitational field of the sun, and the decrease in the frequency of light emitted in a gravitational field (gravitational redshift). Einstein had already deduced the last two phenomena from the equivalence principle. And whereas he could use the equivalence principle also to give the correct quantitative value for the gravitational redshift, he had to wait until completing the full theory before he could derive the correct angle of the bending of light. Einstein examines these phenomena in a weak spherically symmetric gravitational field of a point mass representing the sun, corresponding to the Newtonian limit of the gravitational field equations in vacuum (eq. 47).

> For Einstein, the requirement that an acceptable theory of gravitation would reduce to the Newtonian theory as a limiting or special case was not only natural but absolutely essential. After all, the knowledge about gravitation contained in classical Newtonian theory is empirically well founded. In the course of the Einstein-Grossman collaboration in the search for a generally covariant theory, this requirement served not only as a condition for an acceptable gravitational field equation but also as a starting point for its construction. In 1912, they rejected the Ricci tensor, a natural candidate for the gravitation tensor in the source-free case, because they thought (erroneously) that it did not reduce to the Newtonian expression in the case of an infinitely weak field. It was only in November 1915, when Einstein was working on the Mercury perihelion problem, that he realized how the Newtonian limit should be interpreted.

The gravitational potential is represented by the metric tensor $g_{\mu\nu}$ (p. 53 [8]). In the special theory of relativity, which completely neglects gravitation, this reduces to the form given in eq. (4). We might expect a weak gravitational field to be represented by a matrix of $g_{\mu\nu}$ that differs from this form by quantities much smaller than 1. Because the gravitational potential in Newton's theory is represented by a single function, we could expect that in the Newtonian limit only the component $g_{44}$, which represents the gravitational potential in that limit, would differ from 1. However, this cannot be the case. The coordinate condition $g = -1$ (in the present case $g$ is the product of the elements in the diagonal) implies that if $g_{44}$ differs from 1, then the other diagonal terms have to be different from −1. This result came as a surprise to Einstein and Grossman. The new theory does not reduce to the Newtonian expression in the limiting case of a weak static gravitational field. Einstein repeated the issue of the Newtonian limit again and again. The misunderstanding of this point until November 1915 was one of the main stumbling blocks in his search for a generally covariant theory.

(40a)

(Schlussbemerkung zum Abschnitt D)

Wir haben nun die allgemeinsten Gesetze abgeleitet, welchen das Gravitationsfeld und die Materie genügen, indem wir uns konsequent eines Koordinatensystems bedienten, für welches $\sqrt{-g} = 1$ wird. Wir erzielten dadurch eine erhebliche Vereinfachung der Formeln und Rechnungen, ohne dass wir auf die Forderung der allgemeinen Kovarianz verzichtet hätten. Denn wir fanden unsere Gleichungen durch Spezialisierung (des Koordinatensystems aus) allgemein kovarianten Gleichungen.

Immerhin ist die Frage nicht ohne formales Interesse, ob bei entsprechender verallgemeinerter Definition der Energiekomponenten des Gravitationsfeldes und der Materie auch ohne Spezialisierung des Koordinatensystems Erhaltungssätze von der Gestalt der Gleichung (56) sowie Feldgleichungen der Gravitation von der Art der Gleichungen (52) bezw. (52a) gelten, derart, dass links eine Divergenz (im gewöhnlichen Sinne), rechts die Summe der Energiekomponenten der Materie und der Gravitation steht. Ich habe gefunden, dass beides in der That der Fall ist. Doch glaube ich, dass sich eine Mitteilung meiner ziemlich umfangreichen Betrachtungen über diesen Gegenstand nicht lohnen würde, da doch etwas sachlich Neues dabei nicht herauskommt

### What did Einstein wish to clarify and emphasize as an afterthought?

Einstein finished the manuscript and numbered the pages and then decided to add a few remarks at the end of part D. The asterisk is the editor's instruction to the typesetter to insert this page at the place marked by the asterisk on the previous page.

Einstein wished to emphasize again that the derivation of the field equation of gravitation and the formulation of the conservation laws are based on a specific choice of coordinates, corresponding to $g = -1$, which simplifies the mathematical expressions but does not affect the generality of the results.

He essentially repeats his final remark at the end of part B (pp. 90–92 [27–28]), where he stated that all the relations in this paper will be presented in the simplified form brought about by this choice of coordinates and added: "It will be an easy matter to revert to the *generally* covariant equations, if this seems desirable in a special case."

We know that at the time of writing this manuscript, he considered reformulating the derivation of the field equations in arbitrary coordinates. This becomes clear from the five-page manuscript presented in this book, which he initially intended to include in the body of this article and later as an addendum. Finally, he decided not to include it here and published it about half a year later as a separate article: "The Hamiltonian Principle and the General Theory of Relativity." (The English translation of this article is given in the appendix.)

It is possible that the remarks on this page are a substitute for that addendum and that the last words provide an explanation: "I do not think that the communication of my somewhat extensive reflections on this subject would be worthwhile, because after all they do not give us anything that is materially new."

EINSTEIN: HAMILTONsches Prinzip und allgemeine Relativitätstheorie   1111

# HAMILTONsches Prinzip und allgemeine Relativitäts- theorie.

## Von A. EINSTEIN.

beliebige Geschwindigkeiten $v = \sqrt{\frac{dx_1^2}{dx_4} + \frac{dx_2^2}{dx_4} + \frac{dx_3^2}{dx_4}}$ auftreten können, die kleiner sind als die Vakuum-Lichtgeschwindigkeit ($v < 1$). Will man sich auf den fast ausschliesslich der Erfahrung sich darbietenden Fall beschränken, dass $v$ gegen die Lichtgeschwindigkeit klein ist, so bedeutet dies, dass die Komponenten $\left(\frac{dx_1}{ds}, \frac{dx_2}{ds}, \frac{dx_3}{ds}\right)$ als kleine Grössen

$$\frac{dx_1}{ds} \ll 1$$

zu behandeln sind, während $\frac{dx_4}{ds}$ bis auf Grössen zweiter Ordnung gleich 1 setzt (zweiter Gesichtspunkt der Approximation).

Nun bewerkenwir, dass nach dem ersten Gesichtspunkte der Approximation die Grössen $\Gamma_{\mu\nu}^\tau$ alle kleine Grössen mindestens erster Ordnung in dieser Gleichung sind. Ein Blick auf (46) lehrt also, dass nach dem zweiten Gesichtspunkt der Approximation nur die Glieder zu berücksichtigen sind, für welche $\mu = \nu = 4$ ist. Bei Beschränkung auf Glieder niedrigster Ordnung erhält man anstelle von (46) zunächst die Gleichungen

$$\frac{d^2 x_\tau}{dt^2} = \Gamma_{44}^\tau,$$

wobei $ds = dx_4 = dt$ gesetzt ist, oder unter Beschränkung auf Glieder, die nach dem ersten Gesichtspunkte der Approximation erster Ordnung sind:

$$\frac{d^2 x_\tau}{dt^2} = \begin{bmatrix} 4 & 4 \\ & \tau \end{bmatrix} + \begin{bmatrix} 4 & 4 \\ & 2 \end{bmatrix} + \begin{bmatrix} 4 & 4 \\ & 3 \end{bmatrix} \begin{bmatrix} 4 & 4 \\ & 4 \end{bmatrix}$$

$$\frac{d^2 x_\tau}{dt^2} = \begin{bmatrix} 4 & 4 \\ & \tau \end{bmatrix} \quad (\tau = 1, 2, 3)$$

$$\frac{d^2 x_4}{dt^2} = -\begin{bmatrix} 4 & 4 \\ & 4 \end{bmatrix}$$

Setzt man ausserdem voraus, dass das Gravitationsfeld ein quasistatisches sei, indem man sich auf den Fall beschränkt, dass das das Gravitationsfeld erzeugende Materie nur langsam bewegt, so kann man auf der rechten Seite Ableitungen nach der Zeit neben solchen nach den örtlichen Koordinaten vernachlässigen, sodass man erhält

$$\frac{d^2 x_\tau}{dt^2} = -\frac{1}{2} \frac{\partial g_{44}}{\partial x_\tau} \quad (\tau = 1, 2, 3) \quad \cdot (62)$$

Dies ist die Bewegungsgleichung des materiellen Punktes nach Newtons Theorie, wobei $\frac{g_{44}}{2}$ die Rolle des Gravitationspotentiales spielt. Das Merkwürdige an diesem Resultat ist, nur die Komponente $g_{44}$ des Fundamentaltensors allein in erster Näherung auf die Bewegung des materiellen Punktes bestimmt.

Wir wenden uns nun zu den Feldgleichungen (53). Dabei ist zu berücksichtigen, dass der Energietensor der „Materie" fast ausschliesslich durch die Dichte $\rho$ der Materie im engeren Sinne bestimmt wird, d. h. durch das zweite Glied der rechten Seite von (58) (bezw. (58a) oder (58b)). Bildet man die uns interessierende Näherung, so verschwinden alle Komponenten bis auf die Komponente

$$T_{44} = \rho = T.$$

## What does the metric tensor look like in the Newtonian limit?

Einstein now applies an approximation procedure to reduce the equation of motion of a material particle in a gravitational field (eq. 46) to its Newtonian limit. The right-hand side contains derivatives of the space and time coordinates with respect to the time of motion along the particle trajectory. The ones with $\mu, v = 1,2,3$ correspond to material velocities, which, in the Newtonian limit, are much smaller than the velocity of light (smaller than 1 in this notation) and can be ignored. This leads to the conclusion that only the term with $\mu = 4$, $v = 4$ survives and to eq. (67). This is the equation of motion of a material point in Newton's theory. Einstein notes: "What is remarkable in this result is that the component $g_{44}$ of the fundamental tensor alone defines, to the first approximation, the motion of the material point." Other components of the metric tensor are still dependent on the position in spacetime, indicating that the curvature of spacetime is preserved to the first approximation. These components, however, do not affect the motion of a material point. Einstein then applies the same approximation to the field equations (53) and derives eq. (68) (on the next page), which is exactly Newton's equation for the gravitational potential generated by a mass density $\rho$.

In December 1915, Einstein wrote to his friend Michele Besso about his new theory: "Most gratifying is the agreement with perihelion motion and the general covariance; strangest, however, is the circumstance that Newton's theory of the field is incorrect already in the first order. It is just the circumstance that the $g_{11}$, $g_{22}$, $g_{33}$ (components of the metric tensor) do not appear in first-order approximation of the equation of motion which determines the simplicity of Newton's theory."

## Why was Einstein pleasantly surprised?

On December 22, 1915, Einstein received a letter from the astrophysicist Karl Schwarzschild, who was writing from the Russian front. Schwarzschild informed Einstein that he had solved completely the problem posed in the paper on the Mercury perihelion problem. This was the first exact solution of the field equations of general relativity for a single spherical nonrotating mass. Einstein treated this problem in Cartesian coordinates and derived an approximate solution. In his derivation, Schwarzschild used a more convenient system of coordinates. He wrote: "It is a wonderful thing that the explanation of the Mercury problem emerges so convincingly from such an abstract idea," and he ended his letter saying: "As you see, the war is kindly disposed of me, allowing me, despite fierce gunfire at a decidedly terrestrial distance, to take this walk into this your land of ideas." Einstein urged him to publish his result and promised that he himself would report on it at the next meeting of the Prussian Academy of Sciences. Einstein responded to Schwarzschild's letter: "I would not have expected that the exact solution of the problem could be formulated so simply. The mathematical treatment of the subject appeals to me exceedingly."

Schwarzschild died in May 1916 at the age of 43. Einstein wrote an obituary in the reports of the Prussian Academy of Sciences praising his work and achievements. The Schwarzschild solution eventually became the basis of modern research on black holes and a cornerstone in the study of astrophysical implications of the general theory of relativity.

(42)

Auf der linken Seite von (53) ist das zweite Glied (von zweiter Ordnung) klein, das erste liefert in der uns interessierenden Näherung

$$+\frac{\partial}{\partial x_1}\left[\begin{matrix}\mu\nu\\1\end{matrix}\right]+\frac{\partial}{\partial x_2}\left[\begin{matrix}\mu\nu\\2\end{matrix}\right]+\frac{\partial}{\partial x_3}\left[\begin{matrix}\mu\nu\\3\end{matrix}\right]-\frac{\partial}{\partial x_4}\left[\begin{matrix}\mu\nu\\4\end{matrix}\right]$$

Das liefert für $\mu = \nu = 4$ bei Weglassung von nach der Zeit differenzierten Gliedern

$$-\frac{1}{2}\left(\frac{\partial^2 g_{44}}{\partial x_1^2}+\frac{\partial^2 g_{44}}{\partial x_2^2}+\frac{\partial^2 g_{44}}{\partial x_3^2}\right)=-\frac{1}{2}\Delta g_{44}$$

Die letzte der Gleichungen (53) liefert also

$$\Delta g_{44}=\kappa\varrho.\quad \ldots\ldots (68)$$

Die Gleichungen (62) und (68) zusammen sind äquivalent dem Newton'schen Gravitationsgesetz.

Für das Gravitationspotential ergibt sich nach (62) und (68) der Ausdruck

$$-\frac{\kappa}{8\pi}\int\frac{\varrho\, d\tau}{r}\quad\ldots\ldots(68a)$$

während Newtons Theorie bei der von uns gewählten Zeiteinheit

$$-\frac{K}{c^2}\int\frac{\varrho\, d\tau}{r}$$

ergibt, wobei $K$ die gewöhnlich als Gravitationskonstante bezeichnete Konstante $6,7\cdot10^{-8}$ bedeutet. Durch Vergleich ergibt sich

$$\kappa=\frac{8\pi K}{c^2}=1,87\cdot10^{-27}\quad\ldots\ldots(69)$$

§ 22. Krümmung der Lichtstrahlen. Verhalten von Massstäben und Uhren im statischen Gravitationsfelde. Perihelbewegung der Planetenbahnen.

Um die Newton'sche Theorie als erste Näherung zu erhalten, brauchten wir von den 10 Komponenten des Gravitationspotentials $g_{\mu\nu}$ nur $g_{44}$ zu berechnen, da nur diese Komponente in die erste Näherung (62) der Bewegungsgleichung des materiellen Punktes im Gravitationsfelde eingeht. Man sieht indessen schon daraus, dass noch andere Komponenten der $g_{\mu\nu}$ von den in (4) angegebenen Werten in erster Näherung abweichen müssen, dass letzteres durch die Bedingung $|g| = -1$ verlangt wird.

Für einen im Koordinatensystems Anfangspunkte des befindlichen Masse erzeugenden Massenpunkt erhält man in erster Näherung die radial symmetrische Lösung

$$\begin{rcases} g_{\varrho\sigma}=-\delta_{\varrho\sigma}-\alpha\frac{x_\varrho x_\sigma}{r^3}\ (\varrho \text{ und } \sigma \text{ zwischen } 1 \text{ und } 3) \\ g_{\varrho4}=g_{4\varrho}=0\ (\varrho \text{ zwischen } 1 \text{ und } 3) \\ g_{44}=1-\frac{\alpha}{r}. \end{rcases}(70)$$

$\delta_{\varrho\sigma}$ ist dabei 1 bezw. 0, je nachdem $\varrho=\sigma$ oder $\varrho\neq\sigma$, $r$ ist die Grösse $+\sqrt{x_1^2+x_2^2+x_3^2}$. Dabei ist wegen (68a)

## How could astronomers help confirm certain predictions of the theory?

By now, Einstein has shown that the basic equations of his relativistic theory of gravitation, the equation of motion (eq. 46), and the field equation (eq. 53) reduce to the classical Newtonian theory in the limit of a weak static gravitational field. Yet, even in this limit, $g_{11}$, $g_{22}$, and $g_{33}$ do not reduce to $-1$, as one could expect. The approximate values of $g_{\mu\nu}$ for a weak static spherically symmetric gravitational field are given in eq. (70). In the last chapter, Einstein examines the effect of this conclusion on the geometric properties of spacetime and uses the results to derive the implied astronomical predictions.

When Einstein realized that he was unfamiliar with the mathematical methods, he turned to the mathematician Grossmann for help. Now he appealed to astronomers for help and advice. Specific predictions of the general theory of relativity that can be confirmed or rejected by direct observations are very important in distinguishing it from alternative theories of gravitation. Einstein had been working on such predictions since 1911. He predicted two previously unknown effects that could help test the general theory of relativity. The first is the deflection of light in a gravitational field. In 1913, Einstein wrote to the astronomer George Hale asking for his advice on the possibility of measuring the deflection of light in the vicinity of the solar rim. Hale's reply was that the only opportunity for detecting this effect would be during a solar eclipse. In 1914, after the breakout of World War I, a German expedition was planned to observe this effect during a solar eclipse in Ukraine, but it was interned for a brief period by the Russian authorities. Note that the predicted angle of deflection was then by a factor of 2 smaller than the correct value predicted by general relativity, which would be derived later.

This item is reproduced by the permission of The Huntington Library, San Marino, California.

The second prediction is the change in color of light in a gravitational field, the so-called gravitational redshift. Einstein tried, with little success, to convince German astronomers to initiate research on these effects. He had early support from Erwin Freundlich, at that time an assistant at the Potsdam observatory, and from the astrophysicist Karl Schwarzschild. He particularly acknowledged the role of Freundlich. In April 1914, he wrote to Ehrenfest: "Astronomer Freundlich has found a method to establish light refraction by Jupiter's gravitational field. In addition, he has established with astounding accuracy the shift of intensity centers of solar lines toward the red. . . ." It turned out, however, that the claim of confirmation of the redshift effect was highly premature. Einstein praised Freundlich in a letter to Schwarzschild (January 1916): "He was the first astronomer to understand the significance of the general theory of relativity and to address enthusiastically the astronomical issues attached to it."

$$\alpha = \frac{\kappa M}{8\pi} \quad \ldots \ldots (20a)$$

wenn mit $M$ die felderzeugende Masse bezeichnet wird. Dass durch diese Lösung die Feldgleichungen (ausserhalb der Masse) in erster Näherung erfüllt werden, ist leicht zu verifizieren.

Wir untersuchen nun die Beeinflussung, welche die metrischen Eigenschaften des Raumes durch das Feld der Masse $M$ erfahren. Stets gilt zwischen den „lokal" (§4) gemessenen Längen und Zeiten (ds) einerseits und den Koordinatendifferenzen $dx_\nu$ andererseits die Beziehung

$$ds^2 = g_{\mu\nu}\, dx_\mu\, dx_\nu.$$

Für einen „parallel" der $X$-Achse gelegten Einheitsmassstab wäre beispielsweise zu setzen

$$ds^2 = -1 \; ; \quad dx_2 = dx_3 = dx_4 = 0,$$

Also

$$-1 = g_{11}\, dx_1^2.$$

Liegt der Einheitsmassstab ausserdem auf der $X$-Achse, so ergibt die erste der Gleichungen (20)

$$g_{11} = -\left(1 + \frac{\alpha}{r}\right)$$

Aus beiden Relationen folgt in erster Näherung genau

$$dx = 1 - \frac{\alpha}{2r} \quad \ldots \ldots (21)$$

Der Einheitsmassstab erscheint also mit Bezug auf das Koordinatensystem in dem gefundenen Betrage durch das Vorhandensein des Gravitationsfeldes verkürzt, wenn er radial angelegt wird.

Analog erhält man seine Koordinatenlänge in tangentialer Richtung, indem man beispielsweise setzt

$$ds^2 = -1, \; dx_1 = dx_3 = dx_4 = 0; \; x_1 = r, \; x_2 = x_3 = 0.$$

Es ergibt sich

$$-1 = g_{22}\, dx_2^2 = -dx_2^2 \ldots - (21a)$$

Bei tangentialer Stellung hat also das Gravitationsfeld des Massenpunktes keinen Einfluss auf die Stablänge. Es gilt also die Euklidische Geometrie im Gravitationsfelde nicht einmal in erster Näherung, falls man einen und denselben Stab unabhängig von seinem Ort und seiner Orientierung als Realisierung derselben Strecke auffassen will. Allerdings zeigt ein Blick auf (20a) und (69), dass die zu erwartenden Abweichungen viel zu gering sind um sich bei der Vermessung der Erdoberfläche bemerkbar machen zu können.

Es werde ferner die auf die Zeitkoordinate untersuchte Ganggeschwindigkeit einer Einheitsuhr untersucht, welche in einem statischen Gravitationsfelde ruhend angeordnet ist. Hier gilt für eine Uhrperiode

$$ds = 1, \; dx_1 = dx_2 = dx_3 = 0.$$

## What is the length of rods and the pace of clocks in a gravitational field?

We already know that the length of rods and the pace of clocks depend on the gravitational field at their location in spacetime. Einstein now demonstrates this dependence in a gravitational field produced by a spherically symmetric mass.

Using the values of $g_{\mu\nu}$ of eq. 70 (on the previous page), Einstein examines the effect of the gravitational field produced by a mass at the origin of coordinates on the observed length of rods. The conclusion is that a rod located along the radius is slightly shortened (eq. 71), while the length of a rod laid in the perpendicular direction is not affected by the gravitational field. Consequently, if such a rod is used to measure the diameter and the circumference of a circle around the origin of coordinates, the ratio of the circumference to the diameter is found to be different from $\pi$. Thus Euclidean geometry does not hold even to a first approximation in the gravitational field.

Einstein proceeds to examine the effect of such a gravitational field on the pace of clocks (on the next page).

## Is there a "viable" alternative theory to general relativity?

At this stage, toward the end of the manuscript, when we are about to discuss the experimental tests of Einstein's theory, it is worth mentioning the gravitational theory of Gunnar Nordström, published in 1912. Nordström's theory is based on a single scalar gravitational potential and is embedded in the special theory of relativity. Einstein had several objections against scalar theories of gravitation, which he discussed with Nordström during the latter's visit to Zurich in June 1913. Following that visit, Nordström published a new version of his theory, which was extensively discussed by Einstein in a lecture he gave in Vienna in September 1913: "On the Present State of the Problem of Gravitation." Despite his dissatisfaction with the fact that Nordström's theory could not explain the inertia of a body with the gravitational effect of the surrounding mass distribution (Mach's principle), Einstein concluded: "In sum, we can say that Nordström's scalar theory, which adheres to the postulate of the constancy of the velocity of light, satisfies all of the conditions that can be imposed on a theory of gravitation given the current state of empirical knowledge." Indeed, Einstein considered this theory the only viable alternative to his and Grossmann's *Entwurf* theory.

In Nordström's theory, there is no deflection of light by a gravitational field; however, at that time a rejection of one or the other theory on empirical grounds was impossible. It was hoped that the planned astronomical observations during the solar eclipse in 1914 would do so. This did not happen at that time (p. 55), but the verdict about the shortcoming of Nordström's theory became clear even before the next solar eclipse in 1919. A calculation of the Mercury perihelion shift based on this theory predicts a retrogression of 7″ (arc seconds), while Einstein's theory predicts the observed progression of 43″ (p. 129 [45]).

(44)

Also ist

$$1 = g_{44}\, dx_4^2,$$

$$dx_4 = \frac{1}{\sqrt{g_{44}}} = \frac{1}{\sqrt{1+(g_{44}-1)}} = 1 - \frac{g_{44}-1}{2}$$

oder

$$dx_4 = 1 + \frac{\kappa}{8\pi}\int \frac{\varrho\, d\tau}{r} \quad \dots (72)$$

Die Uhr läuft also langsamer, wenn sie ~~sich~~ in der Nähe ponderabler der Tentralstern Massen aufgestellt ist. Es folgt daraus, dass ~~von~~ der Oberfläche grosser Sterne zu ~~uns~~ gelangenden Lichtes nach dem roten Spektralende verschoben erscheinen müssen.

Wir untersuchen ferner den Gang der Lichtstrahlen im statischen Gravitationsfeld. Gemäss der speziellen Relativitätstheorie ist die Lichtgeschwindigkeit durch die Gleichung

$$-dx_1^2 - dx_2^2 - dx_3^2 + dx_4^2 = 0$$

gegeben, also gemäss der allgemeinen Relativitätstheorie durch die Gleichung

$$ds^2 = g_{\mu\nu}\, dx_\mu dx_\nu = 0 \quad \dots\dots (73)$$

Ist die Richtung, d. h. das Verhältnis $dx_1 : dx_2 : dx_3$ gegeben, so liefert die Gleichung (73) die Grössen $\frac{dx_1}{dx_4}, \frac{dx_2}{dx_4}, \frac{dx_3}{dx_4}$, und somit die Geschwindigkeit

$$\sqrt{\left(\frac{dx_1}{dx_4}\right)^2 + \left(\frac{dx_2}{dx_4}\right)^2 + \left(\frac{dx_3}{dx_4}\right)^2} = \gamma,$$

im Sinne der Euklidischen Geometrie definiert. Man erkennt leicht, dass die Lichtstrahlen gekrümmt verlaufen müssen mit Bezug auf das Koordinatensystem, falls die $\gamma$ nicht konstant sind. Ist $n$ eine Richtung senkrecht zur Lichtfortpflanzung, so ergibt das Huygens'sche Prinzip, dass der Lichtstrahl (in der Ebene $(\gamma, n)$ betrachtet) die Krümmung $-\frac{\partial \gamma}{\partial n}$ besitzt.

Wir untersuchen die Krümmung, welche ein Lichtstrahl erleidet, welche an Lichtstrahl erleidet, der im Abstand $\Delta$ an einer Masse $M$ vorbeigeht. Wählt man das Koordinatensystem gemäss der nebenstehenden Skizze, so ist die gesamte Biegung $B$ des Lichtstrahles (positiv gerechnet, wenn sie nach dem Ursprung hin konkav ist) in genügender Näherung gegeben durch

$$B = \int_{-\infty}^{+\infty} \frac{\partial \gamma}{\partial x_1}\, dx_2,$$

während (73) und (70) ergeben

$$\gamma = \sqrt{\frac{g_{44}}{g_{22}}} = 1 + \frac{\alpha}{2r}\left(1 + \frac{x_2^2}{r^2}\right)$$

Figur

Für das Bestehen eines derartigen Effektes sprechen nach E. Freundlich spektrale Beobachtungen an Fixsternen bestimmter Typen. Eine endgültige Prüfung dieser Konsequenz steht indes noch aus.

### What observation catapulted Einstein to world celebrity status?

The first result on this page is that a clock moves more slowly in the neighborhood of a massive body. Equation (72) allows us to estimate the size of this effect. An atom emitting light may be viewed as a clock. The slowing down of such "atomic clocks" in a gravitational field means that the frequency of these oscillations, and hence of the emitted light, is reduced. The "color" of light of lower frequency is shifted to the red end of the spectrum of light. In the footnote, Einstein acknowledges Freundlich for having observed such an effect from spectroscopic observations on certain stars, remarking that this is not yet a crucial test.

Next, Einstein examines the deflection undergone by a ray of light passing near a mass $M$ at a distance $\Delta$ and concludes that a ray of light going past the sun undergoes a deflection of 1.7″ (this result appears on the next page).

The predicted angle of bending was twice the value that Einstein obtained from the equivalence principle, which appears in his letter to George Hale (p. 123). This value also could have been explained by an old (almost forgotten) Newtonian theory of light. Einstein's prediction was confirmed by astronomical observations during the solar eclipse of 1919, performed by a British expedition headed by the astronomer Arthur Eddington, and Einstein became a world celebrity overnight. It was not just the phenomenon itself but also the measured angle that caused a sensational title in the *London Times* on November 7, 1919.

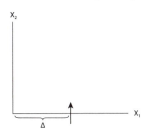

The redshift result was not confirmed till the late 1950s. Nowadays, the effect of gravitation on the rate of clocks has to be taken into account in the timing system of GPS technology.

A square is cut out of this page, reflecting the usual editorial procedure in those days: the editor removed the figures from the manuscript and sent them to the graphics department; the rest was sent to typesetting. The missing figure is reproduced here. Einstein used this diagram to explain his choice of the coordinates $x_1$ and $x_2$ that appear in the calculation.

Die Ausrechnung ergibt

$$B = \frac{2\alpha}{\Delta} = \frac{\kappa M}{4\pi\Delta} \quad \ldots \ldots (74)$$

Ein an der Sonne vorbeigehender Lichtstrahl erfährt demnach eine Biegung von 1,7″, ein am Jupiter (Planeten) vorbeigehender eine solche von etwa 0,02″.

Berechnet man das Gravitationsfeld um eine Grössenordnung genauer und ebenso mit entsprechender Genauigkeit die Bahnbewegung eines materiellen Punktes von relativ unendlich kleiner Masse, so erhält man gegenüber den Kepler–Newton'schen Gesetzen der Planetenbewegung eine Abweichung von folgender Art. Die grosse Achse (Bahnellipse) eines Planeten erfährt in Richtung der Bahnbewegung eine langsame Drehung, vom Betrage

$$\varepsilon = 24\pi^3 \frac{a^2}{T'^2 c^2 (1-e^2)} \quad \ldots \ldots (75)$$

pro Umlauf. In dieser Formel bedeutet $a$ die grosse Halbachse, $c$ die Lichtgeschwindigkeit in üblichem Masse, $e$ die Exzentrizität, $T$ die Umlaufszeit in Sekunden.

Die Rechnung ergibt für den Planeten Merkur eine Drehung der Bahn von 43″ pro Jahrhundert, genau entsprechend der Konstatierung der Astronomen (Leverrier); diese fanden nämlich einen durch Störungen der übrigen Planeten nicht erklärbaren Rest der Perihelbewegung dieses Planeten von der angegebenen Grösse.

$$(\varepsilon \ 20/3 \ 16)$$

Bezüglich der Rechnung verweise ich auf die Originalabhandlungen
A. Einstein. Sitz. Ber. d. Preuss. Akad. d. W. XLVII. 1915. S. 831
K Schwarzschild. Sitz Ber d. Preuss. Akad. d. W. VII. 1916. S. 189.

## Explanation of the motion of perihelion of planet Mercury: From disappointment to triumph

Newton's theory of gravitation confirms Kepler's observation that the planets move around the sun in elliptic orbits. If there were only one planet in the solar system, then the position of the perihelion of the orbit (the point of closest approach to the sun) would be fixed in space. However, owing to the influence of the other planets, there is a slow precession of the perihelion. Astronomers discovered that the orbit of the planet Mercury around the sun, from the point of view of Earth, rotates by 5600″, or 1.55° (degrees), in 100 years. Most of this rotation could be explained by forces exerted by the other planets, but 43″ was unexplained. This problem had already arisen in 1859, following the work of the French astronomer Urbain Le Verrier, and remained unsolved until Einstein's general theory of relativity.

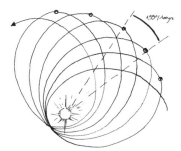

In December 1907, when Einstein was taking his initial steps on the way to the relativistic theory of gravitation, even before he had any kind of a theory, he realized that it might provide an answer to the longstanding problem. He wrote to his friend Conrad Habicht: "At the moment I am working on a relativistic analysis of the unexplained secular changes in the perihelion of Mercury." Shortly after the publication of the Einstein-Grossmann *Entwurf* theory, Einstein and his friend Michele Besso performed a calculation to test whether the new theory could account for the anomalous precession of the perihelion of Mercury. They concluded that only 18″ of the discrepancy could be explained by the new theory. This disappointing result did not cause Einstein to doubt the validity of the *Entwurf* theory. In fact, he never published this result and ignored it completely until November 1915, when he used it as an argument to justify the decision to abandon the *Entwurf* theory.

The last page presents the result that he obtained when still working on the final stages of his new theory, which he presented on November 18, 1915, to the Prussian Academy of Sciences. This result remained unchanged in the final version of the theory as well. In the footnote, Einstein refers the reader to this work and to the work of Schwarzschild for the details of the calculation. The last sentence of the manuscript reads: "Calculation gives for the planet Mercury a rotation of the orbit of 43″ per century, corresponding exactly to astronomical observations (Leverrier); for astronomers have discovered in the motion of perihelion of this planet, after allowing for disturbances by other planets, an inexplicable remainder of this magnitude."

The editor's pencil remark in the lower part of the page indicates the date this manuscript was received, March 20, 1916.

Anhang: Darstellung der Theorie ausgehend von einem — (1)
Variationsprinzip.          A. Einstein.

§1. Die Feldgleichungen der Gravitation und der Materie.

Wir setzen voraus, dass sich die
~~Die~~ Feldgleichungen der Gravitation und die aller anderen
Vorgänge lassen sich mit Vorteil aus einem allgemeinen Variationssatze
~~nach~~ von Hamilton'scher Form ableiten

$$\delta \left\{ \int \mathfrak{H} \, d\tau \right\} = 0 \quad \cdots \quad (76)$$

wobei $d\tau = dx_1 \, dx_2 \, dx_3 \, dx_4$ gesetzt ist,
ableiten. $\mathfrak{H}$ sei dabei eine Funktion der $g^{\mu\nu}$ und $g_\sigma^{\mu\nu} \left( = \dfrac{\partial g^{\mu\nu}}{\partial x_\sigma} \right)$ und ferner
gewisser Funktionen $q_{(\rho)}$ und ihrer Ableitungen nach den $x_\nu$, welche
die materiellen (Zustände und Vorgänge im weitesten Sinne beschreiben. Unter
diesen verstehen wir alle Zustände und Vorgänge exklusive derjenigen, welche
allein das Gravitationsfeld betreffen, also ausser den Bewegungen und
Zustandsänderungen der Materie im engeren Sinne auch die elektromagnetischen Vorgänge im Vakuum. Bei der Variation sollen ~~sich~~ die
~~von einander unabhängigen~~ $g^{\mu\nu}$ und $q_{(\rho)}$ unabhängig voneinander
variiert werden, wobei die $\delta g^{\mu\nu}$, $\delta q_{(\rho)}$ und eventuell gewisse Ableitungen
dieser Variationen $\delta q$ an den Integrationsgrenzen verschwinden sollen.

Durch das Einklammern des Index $\rho$ bei $q_{(\rho)}$ soll
angedeutet werden, dass die Stellung dieses Index über den transfor-
mationscharakter und die Anzahl der zur Beschreibung der
"Materie" zu verwendenden Funktionen nichts aussagen soll. Diese
Unbestimmtheit der Darstellung erscheint mir vorläufig nötig,
~~weil wir über die endgültige~~ theoretische Darstellung der Materie noch
recht wenig ~~sagen~~ wissen. ×

Um der Superponierbarkeit der Gravitationsfelder und
der die Materie bildenden ~~Kom~~ Felder zu entsprechen, nehmen wir
ferner an, dass $\mathfrak{H}$ sich als Summe in der Form

$$\mathfrak{H} = \mathfrak{H}^{(g)} + \mathfrak{H}^{(m)} \qquad \mathfrak{H} = \mathfrak{G} + \mathfrak{M} \quad \cdots \quad (77)$$

darstellen lasse, derart, dass $\mathfrak{G}$ nur von den $g^{\mu\nu}$ und $g_\sigma^{\mu\nu}$, $\mathfrak{M}$ nur von
den $g^{\mu\nu}$, $q_{(\rho)}$ und den Ableitungen abhänge. Man erhält dann aus
(77) durch Variieren nach den $g^{\mu\nu}$ die Gleichungen ~~der Gravitation~~

$$\frac{\partial}{\partial x_\alpha} \left( \frac{\partial \mathfrak{G}}{\partial g_\alpha^{\mu\nu}} \right) - \frac{\partial \mathfrak{G}}{\partial g^{\mu\nu}} = \frac{\partial \mathfrak{M}}{\partial g^{\mu\nu}} \quad \cdots \quad (78)$$

durch Variieren nach den $q_{(\rho)}$ die Gleichungen

× Die ~~von~~ Hilbert im Anschluss an Mie eingeführte Voraussetzung, dass sich
die ~~Materie~~ durch ~~einen~~ Vierervektor $q_\rho$ und dessen erste Ableitungen darstellen
lasse, halte ich für wenig aussichtsvoll.

### Why did Einstein decide not to include this "Appendix" in the printed version of the manuscript "Foundation of General Relativity"?

This is the first of five pages that Einstein initially intended to include in the manuscript. Judging by the "§14" (between the title lines) and the equation numbers on this page, which he crossed out, he planned to insert these pages after section 13 and to proceed immediately to the derivation of the gravitational field equations from the "variational principle" (explained on p. 97). Einstein then decided to do it more gradually. First, with the help of the variational principle, he derived the field equation in the absence of matter and then introduced matter in the same way that the energy and momentum of the gravitational field appeared in that equation. Einstein then intended to attach this calculation as an appendix to the manuscript, stating this explicitly in the title and numbering the pages accordingly. In the end, he did not include it at all. In October 1916, he submitted to the Royal Prussian Academy an article titled "Hamilton's Principle and the General Theory of Relativity" (in our remarks, we shall refer to this article as the "October paper"), which bears significant similarities to and significant differences from the appendix to our manuscript, which was never published.

> The derivation of the gravitational field equation in part C of the manuscript depends on the coordinate condition $-g = 1$. On page 40a, Einstein states that it is possible to formulate the gravitational field equations and the energy-momentum conservation law in the form achieved at the end of part C, even without choosing specific systems of coordinates. This appendix would have demonstrated that. However, Einstein eventually came to the conclusion that this would not be worthwhile because one would not learn anything new. He therefore decided not to include it in his "Foundation of General Relativity" review article.

### Einstein applies a Hamiltonian (Lagrangian) formulation—different from Hilbert's and different from his own previous one

Einstein applies the variational method to a Hamiltonian function—which today we would call a Lagrangian—which depends on the components of the metric tensor (the gravitational potentials), on their derivatives, and on the parameters describing matter and their derivatives (everything outside of the gravitational field). Einstein does not specify the nature of these parameters contrary to Hilbert, who, addressing the same problem, assumed that these parameters are the four components of the electromagnetic potential. Hilbert's approach was based on Mie's theory (p. 99), which stipulates that all matter is of electromagnetic origin and that the energy-momentum tensor of matter should depend only on electromagnetic quantities. In the footnote on this page, Einstein asserts that Hilbert's assumption that the Hamiltonian depends only on such quantities and their derivatives is not very promising.

Einstein separates the Hamiltonian into two parts, $\mathfrak{G}$ and $\mathfrak{M}$, the first depending on the gravitational field parameters only, and the second depending on all the gravitational and matter variables. He then derives the field equations, satisfied by the two parts of the Hamiltonian, eqs. (78) and (79) (at the top of the next page). The latter appears here as a mathematical expression. Einstein should have equated it to zero to make it the equation for the matter part, $\mathfrak{M}$, of the Hamiltonian.

$$\frac{\partial}{\partial x_\alpha}\left(\frac{\partial \mathfrak{M}}{\partial \frac{\partial q_{(\varrho)}}{\partial x_\alpha}}\right) - \frac{\partial \mathfrak{M}}{\partial q_{(\varrho)}} \quad \cdots \cdots (54)(79)$$

Die letzten Gleichungen (79) nennen wir die Feldgleichungen der Gravitation, die Gleichungen (54) die Feldgleichungen der Materie. Die Gleichungen (50) setzen voraus, dass $\mathfrak{M}$ nur von den ersten Ableitungen der $q_{(\varrho)}$ nach den Koordinaten abhänge; kommen auch höhere Ableitungen der $\mathfrak{M}$ vor, so treten weitere Glieder auf. Unsere nachfolgenden Überlegungen gelten jedoch unabhängig hiervon.

§2. Formale Konsequenzen aus der Forderung der allgemeinen Kovarianz.

Wir stellen nun die dem allgemeinen Relativitätspostulat entsprechende Forderung auf: Die Bedingung (76) und damit auch das aus (78) und (79) bestehende Gleichungssystem soll beliebigen Substitutionen der Raum - Zeit - Koordinaten gegenüber kovariant sein. Diese Forderung lässt sich wegen der Invarianz von $\sqrt{-g}\,d\tau$ dadurch erfüllen, dass man $\frac{\mathfrak{H}}{\sqrt{-g}}$ gleich einer Invarianten setzt. So habe es ist dann ist nämlich das Integral das auf der linken Seite von (76) stehende Integral, und damit auch dessen Variation eine Invariante. Damit jedoch die Gleichung (76) invariante Bedeutung erhalte, ist nicht unbedingt nötig, dass $\frac{\mathfrak{H}}{\sqrt{-g}}$ eine Invariante sei; wir gehen vielmehr wie folgt vor.

Es sei zunächst $\frac{\mathfrak{M}}{\sqrt{-g}}$ eine Invariante. Für die Wahl von $\mathfrak{H}$ dient folgende Erwägung. Aus dem in Gleichung (43) gegebenen Riemann'schen Tensor lässt sich die Invariante

$$K = g^{\mu\nu} B^\iota_{\mu\nu\tau} = g^{\mu\nu}\left[-\frac{\partial}{\partial x_\iota}\left\{{\mu\nu \atop \iota}\right\} + \frac{\partial}{\partial x_\tau}\left\{{\mu\tau \atop \tau}\right\} + \left\{{\mu\sigma \atop \tau}\right\}\left\{{\nu\tau \atop \sigma}\right\} - \left\{{\mu\nu \atop \sigma}\right\}\left\{{\sigma\tau \atop \tau}\right\}\right].$$ (80)

bilden. Die Mathematiker haben bewiesen, dass dies die einzige Invariante ist, welche aus den $g^{\mu\nu}$ und den ersten und zweiten Ableitungen der $g^{\mu\nu}$ nach den Koordinaten gebildet werden kann, und welche in den zweiten Ableitungen der $g^{\mu\nu}$ linear ist. Es läge also nahe, die zu wählende Hamilton'sche Funktion $\mathfrak{H}$ für das Gravitationsfeld (bis auf einen konstanten Faktor) gleich die Funktion $K\sqrt{-g}$ zu wählen, da man bei dieser Wahl die Invarianz des Integrals in (76) erhalten wäre. Diese Wahl hätte aber den formalen Nachteil, dass $\mathfrak{H}$ auch von den zweiten Ableitungen der $g^{\mu\nu}$ nach den Koordinaten abhinge, was wir vermeiden wollen. Das Integral

$$\int K\sqrt{-g}\,d\tau$$

—————————
* Hilbert und Lorentz haben zuerst diesen Weg eingeschlagen.

## Why did Einstein decide to publish a modified version of this appendix after all? What was the role of Lorentz and Hilbert?

After completing his general theory of relativity, Einstein became increasingly aware of the important role of Hamilton's formulation, and he corresponded on the topic with colleagues. In January 1916, he wrote to Lorentz: "Only too well do I understand your attempt to derive gravitation from the field equations in the manner of Hamilton's principle. I myself am compelled to derive the Hamiltonian function retroactively, in order to derive the expression for the conservation laws conveniently." In spite of having achieved this derivation, he did not include it in his final review article, submitted two months later. In the same letter, he added: "Nevertheless, I must admit that I actually do not see in Hamilt. Princip. anything more than a means toward reducing a system of tensor equations to a scalar equation for which the conservation laws are always satisfied and easily derived."

On this page, Einstein explains how to assure that the "action" defined by the integral in eq. (76) will be an invariant (a scalar) quantity, so that the "variation" of this action will yield generally covariant equations. In a footnote, he indicates that this way had been proposed by Hilbert and Lorentz. In the October paper, Einstein mentions Hilbert and Lorentz in the opening sentence and, in a footnote, references to their work.

It is possible that the work of Lorentz and Hilbert prompted Einstein to publish his own version of the subject matter. He could have done it half a year earlier. In any case, it was important to him to demonstrate the relation between covariance properties and conservations laws in full generality. It is instructive to quote the first paragraph of the October paper: "H. A. Lorentz and D. Hilbert have recently succeeded in presenting the theory of general relativity in a particularly comprehensive form by deriving its equations from a single variational principle. The same shall be done in this paper. My aim here is to present the fundamental connections as transparently and comprehensively as the principle of general relativity allows. In contrast to Hilbert's presentation, I shall make as few assumptions about the constitution of matter as possible. On the other hand, and in contrast to my own very recent treatment of the subject matter, the choice of coordinates shall be made completely free."

Einstein applies the variational method to the gravitational part of the Hamiltonian, 𝕲. The only appropriate invariant quantity, which depends on the components of the metric tensor and their first and second derivatives, is formed from the metric tensor and the Riemann tensor by internal multiplication and contraction (eq. 80). Today we would call it the *Ricci scalar*. In the October paper, Einstein refers in the parallel paragraph for the first time to the Riemann tensor as the Riemann curvature tensor. Although "spacetime curvature" has become the common concept to describe the effect of massive objects on spacetime, Einstein had not used this term earlier.

In a letter to Hermann Weyl, Einstein is explicitly critical of Hilbert's approach: "Hilbert's assumption about matter appears childish to me in the sense of a child who does not know. At all events, mixing the solid considerations originating from the relativity postulate with such bold unfounded hypotheses about the structure of the electron or matter cannot be sanctioned. I gladly admit that the search for a suitable hypothesis, or for the Hamilton function for the structural makeup of the electron, is one of the most important tasks of theory today. The 'axiomatic method' can be of little use here, though." The last sentence refers to Hilbert's ambition to construct an axiomatic formulation of physics, bringing it closer to a science like geometry.

lässt sich nach (80) als Summe von vier Integralen schreiben, von denen die beiden ersten sich durch partielle Integration umformen lassen. Man erhält unter Verwendung der Gleichungen (29α) und (31) durch

$$-\frac{1}{\kappa}\int K\sqrt{-g}\,d\tau = \int \mathfrak{G}\cdot d\tau + F, \quad \cdots \cdots (81)$$

wobei gesetzt ist

$$\mathfrak{G} = \frac{1}{\kappa}\sqrt{-g}\,g^{\mu\nu}\left[\left\{{}^{\mu\,\alpha}_{\,\beta}\right\}\left\{{}^{\nu\,\beta}_{\,\alpha}\right\} - \left\{{}^{\mu\nu}_{\,\alpha}\right\}\left\{{}^{\alpha\,\beta}_{\,\beta}\right\}\right]. \quad \cdots (82)$$

$\kappa$ bedeutet eine Konstante, F ein über die Begrenzung des betrachteten vierdimensionalen Gebietes erstreckt es Integral, ₓ dessen Integrand eine Funktion der $g^{\mu\nu}$ und $g^{\mu\nu}_\sigma$ ist. Durch diese Wahl von $\mathfrak{H}$ wird zwar nicht die Invarianz des Integrales

$$\int \mathfrak{G}\cdot d\tau$$

erzielt, wohl aber die Invarianz der Variation dieses Integrals, wenn die $\delta g^{\mu\nu}$ so gewählt worden, dass sie samt ihren ersten Ableitungen an der Begrenzung des Integrationsgebietes verschwinden. Es verschwindet nämlich in diesem Falle $\delta F$, sodass man durch Variieren von (52) erhält

$$\delta\left\{\int \mathfrak{G}\,d\tau\right\} = -\frac{1}{\kappa}\,\delta\left\{\int K\sqrt{-g}\,d\tau\right\}$$

Da die Rechte Seite dieser Gleichung wegen der Invarianz von K eine Invariante ist, gilt dasselbe auch von der linken Seite.

Während die Kovarianz Forderung für die Wahl des Hamilton'schen Funktion der Materie noch unübersehbar viele Möglichkeiten offen lässt, liefert sie aber die Hamilton'sche Funktion für das Gravitationsfeld, und damit die linke Seite der Gleichungen (28) beinahe vollständig, ohne jede zusätzliche Voraussetzung.

§ 3. Eigenschaften der Hamilton'schen Funktion $\mathfrak{G}$.

Aus der Thatsache, dass

$$\delta\left\{\int \mathfrak{G}\,d\tau\right\}$$

bei verschwinden der Variationen an den Integrationsgrenzen invariant ist, folgt in bekannter Weise die Invarianz des Integrales

$$\int \delta g^{\mu\nu}\left[\frac{\partial}{\partial x_\alpha}\left(\frac{\partial \mathfrak{G}}{\partial g^{\mu\nu}_\alpha}\right) - \frac{\partial \mathfrak{G}}{\partial g^{\mu\nu}}\right]\,d\tau.$$

Hieraus folgt wegen des Tensorcharakters und der freien Wählbarkeit der $\delta g^{\mu\nu}$, dass auch die linke Seite von (28) ein mit $\sqrt{-g}$ multiplizierter

### Is the conservation principle satisfied without any restrictions?

On this page, Einstein formulates the variation of the gravitational part of the Hamiltonian in a form that makes its invariant (scalar) character apparent. He points out that the choice of the second part of the Hamiltonian, $\mathfrak{M}$, leaves a significant degree of freedom and does not imply any assumptions or restrictions on the gravitational part of the Hamiltonian and, therefore, not on the left-hand side of the gravitational field equation (eq. 78).

He emphasized this point in a letter to Lorentz from November 1916, accompanying the October paper he sent to him on that occasion: "Particularly, I wanted to show that the general relativity concept regarding matter does not limit the variety of possible choices for the Hamiltonian function to a higher degree than the postulate of special relativity, since the conservation laws are satisfied by any choice of $\mathfrak{M}$." He then repeats his criticism of Hilbert: "The choice made by Hilbert thus appears to have no justification."

### 1916: A year of hard work and new beginnings

In December 1915, Einstein wrote to Besso: "The boldest dreams have now been fulfilled. *General* covariance. Mercury's perihelion motion wonderfully precise. . . . from your contented but quite worn-out (*ziemlich katputen*), Albert."

He had good reason to be content and could have relaxed for a while, communicating with friends and colleagues and enjoying the feeling of achievement. But this did not happen. The year 1916 was a year of hard work and new beginnings.

The first article Einstein published after having completed and submitted the manuscript on the foundation of general relativity was an approximate solution of the field equations in a specific coordinate system, suggested to him by the astronomer Willem de Sitter. In this paper, Einstein discussed gravitational waves, concluding that accelerated massive objects generate changes in the metric describing the local properties of spacetime (today we would refer to it as changes in curvature), which propagate, like waves, with the speed of light. However, he committed a calculational error that led to the strange result that "there should exist gravitational waves without energy transport." In 1918, he corrected this error, admitting that his previous treatment of the subject "was marred by a regrettable error in computation." He also derived a famous formula for the energy loss of a system emitting gravitational radiation. Nevertheless, the issue remained controversial. In 1937, Einstein even tried to disprove the existence of gravitational waves.

After much additional work on the issue, physicists today are convinced of the existence of gravitational waves produced by accelerated massive objects. For example, spiraling binary stars have been demonstrated to be a powerful source of such waves. Owing to the astronomical distances to the sources of gravitational waves, their effects on Earth are predicted to be extremely small. In spite of extensive continuing attempts to detect gravitational waves by direct measurements with ever more sensitive detectors, this goal has still not been achieved and remains a challenge at the forefront of research in general relativity.

(4)

kovarianter Tensor ist. Da diese linke Seite nur von den $g^{\mu\nu}, \frac{\partial g^{\mu\nu}}{\partial x_\sigma}, \frac{\partial^2 g^{\mu\nu}}{\partial x_\sigma \partial x_\tau}$, und von den letzteren Grössen linear abhängt, folgt notwendig, dass diese Grösse gleich

$$\pm \sqrt{-g}\left(\alpha B_{\mu\nu} + \beta g_{\mu\nu} g^{\sigma\tau} B_{\sigma\tau}\right)$$

sein muss, wobei $\alpha$ und $\beta$ Konstante bedeuten, und $B_{\mu\nu}$ der in (44) angegebene Ausdruck ist, denn es gibt sonst keine derartige Kovariante. Die Konstanten ergeben sich durch Ausrechnen; es ist

$$\frac{\partial}{\partial x_\alpha}\left(\frac{\partial \mathfrak{H}}{\partial g_\alpha^{\mu\nu}}\right) - \frac{\partial \mathfrak{H}}{\partial g^{\mu\nu}} = \frac{1}{k}\sqrt{-g}\left(B_{\mu\nu} - \frac{1}{2}g_{\mu\nu}K\right)\Bigg\}$$

$$K = g^{\sigma\tau}B_{\sigma\tau}$$

$$(83).$$

Wir leiten ferner zwei identische Gleichungen ab, welche die Hamilton'sche Funktion $\mathfrak{H}$ ~~ungefähren Zusammenhang~~ genügt. Zu diesem Zweck führen wir eine infinitesimale Transformation der Koordinaten durch, indem wir setzen

$$x_\nu' = x_\nu + \Delta x_\nu, \quad \ldots \ldots (84)$$

die $\Delta x_\nu$ sind beliebig wählbare, unendlich kleine Funktionen der Koordinaten. $x_\nu'$ sind die Koordinaten des Weltpunktes im neuen System, dessen Koordinaten im ursprünglichen $x_\nu$ sind. Für jede Grösse oder jede Gruppe von Grössen $\mathfrak{H}$, die bezüglich beliebiger Koordinatensysteme definiert ist, existiert dann ein Transformationsgesetz vom Typus

$$\mathfrak{H}' = \mathfrak{H} + \Delta \mathfrak{H},$$

wobei sich $\Delta \mathfrak{H}$ durch die $\Delta x_\nu$ und deren Ableitungen linear ausdrücken lassen muss. Aus der ~~Kontravarianz der~~ ~~Kovarianz der $g_{\mu\nu}$~~ ~~(Kontravarianz der $g^{\mu\nu}$ folgt mit Rücksicht~~ ~~auf das~~ ~~leicht man~~ ~~mittelst der Gleichung (9)~~ und (84) für die $g^{\mu\nu}$ und $g_\sigma^{\mu\nu}$ die Transformationsgleichungen

$$\Delta g^{\mu\nu} = g^{\mu\alpha}\frac{\partial \Delta x_\nu}{\partial x_\alpha} + g^{\nu\alpha}\frac{\partial \Delta x_\mu}{\partial x_\alpha} \quad \ldots (85)$$

$$\Delta g_\sigma^{\mu\nu} = \frac{\partial \Delta g^{\mu\nu}}{\partial x_\sigma} - g_\alpha^{\mu\nu}\frac{\partial \Delta x_\alpha}{\partial x_\sigma}. \quad \ldots \ldots (86)$$

Da $\mathfrak{H}$ nur von den $g^{\mu\nu}$ und $g_\sigma^{\mu\nu}$ abhängt, ist es möglich mit Hilfe dieser Gleichungen $\Delta \mathfrak{H}$ zu berechnen. Man erhält so die Gleichung

$$\sqrt{-g}\,\Delta\left(\frac{\mathfrak{H}}{\sqrt{-g}}\right) = \mathfrak{H}_\sigma^{\nu}\frac{\partial \Delta x_\sigma}{\partial x_\nu} + 2\frac{\partial \mathfrak{H}}{\partial g_\alpha^{\mu\nu}}g^{\mu\nu}\frac{\partial^2 \Delta x_\sigma}{\partial x_\nu \partial x_\alpha}, \quad \ldots \ldots (87)$$

Having adopted (on the previous page) a specific form of the gravitational Hamiltonian $\mathfrak{G}$, Einstein now expresses the left-hand side of eq. (78) in terms of the contracted Riemann tensor (eq. 83). The tensor $B_{\mu\nu}$ is the tensor $G_{\mu\nu}$ in eq. (44) of the main manuscript. There the condition $-g = 1$ is used to simplify this tensor.

## Einstein acts as a missionary of science

After submitting, in November 1915, the final version of his general theory of relativity, when he was writing the comprehensive summary of the theory with all its ingredients for the scientific community represented by the current manuscript, Einstein was already thinking about writing a popular book on relativity, both the special and the general. In January 1916, he wrote to his friend Besso: "The great success in gravitation pleases me immensely. I am seriously considering writing a book in the near future on special and general relativity theory, although, as with all things that are not supported by a fervent wish, I am having difficulty getting started. But if I do not do so, the theory will not be understood, as simple though it basically is."

Einstein completed the manuscript in December, and the book *Relativity: the Special and the General Theory* was a great success. Between 1917 and 1922, the book appeared in 14 editions in German, and after the confirmation of the bending of light, it also appeared in foreign languages. These editions contained minor textual changes and introductions.

In the introduction to the first edition, Einstein wrote: "The author has spared himself no pains in his endeavor to present the main ideas in the simplest and most intelligible form, and on the whole, in the sequence and connection in which they actually originated." He concluded the introduction with the wish: "May the book bring some one a few happy hours of suggestive thought."

Einstein believed that the laws of nature can be formulated in a number of simple basic principles. This quest for simplicity marked his scientific activity. He also believed that it was his duty to explain these principles in simple terms to the general public and to convey the happiness and satisfaction that the understanding of these principles can generate. This book is one of many examples of Einstein's commitment to his role as a missionary of science.

wobei zur Abkürzung gesetzt ist

$$\mathfrak{S}_\sigma^{\ \nu} = 2\frac{\partial \mathfrak{H}}{\partial g^{\mu\sigma}}g^{\mu\nu} + 2\frac{\partial \mathfrak{H}}{\partial g_\alpha^{\mu\sigma}}g_\alpha^{\mu\nu} + \mathfrak{H}\delta_\sigma^{\ \nu} - \frac{\partial \mathfrak{H}}{\partial g_\nu^{\mu\alpha}}g_\sigma^{\mu\alpha}. \cdots (88)$$

aus (82)

Es ist andererseits leicht zu beweisen, dass $\frac{\mathfrak{H}}{\sqrt{-g}}$ zwar nicht beliebigen Substitutionen, wohl aber linearen Substitutionen gegenüber eine Invariante ist. Hieraus folgt, dass die rechte Seite von (82) stets verschwinden muss, wenn sämtliche $\frac{\partial^2 x_\sigma}{\partial x_\nu \partial x_\alpha}$ verschwinden. Daraus folgt sogleich, dass die Identität

$$\mathfrak{S}_\sigma^{\ \nu} \equiv 0 \cdots (89)$$

bestehen muss.

Wählen wir ferner die $\Delta x_\nu$ so, dass sie unter Wahrung der Stetigkeit in infinitesimaler Nähe der Begrenzung eines betrachteten Gebietes verschwinden, so können wir an Gleichung (81) folgende Betrachtung anknüpfen. Bei der ins Auge gefassten infinitesimalen Substitution ist

$$\Delta \mathcal{F} = 0.$$

Ferner ist wegen der Invarianz von $K$ und $\sqrt{-g}\,d\tau$

$$\Delta\left\{\int K\sqrt{-g}\,d\tau\right\} = 0$$

Es verschwindet also auch

$$\Delta\left\{\int \mathfrak{H}\,d\tau\right\}.$$

Statt die Hieraus folgt wegen der Invarianz von $\sqrt{-g}\,d\tau$ und infolge der Gleichungen (82) und (89) zunächst

$$\int \frac{\partial \mathfrak{H}}{\partial g_\alpha^{\mu\sigma}}g^{\mu\nu}\frac{\partial^2 \Delta x_\sigma}{\partial x_\nu \partial x_\alpha}\,d\tau = 0$$

Formt man diese Gleichung durch zweimalige partielle Integration um, so erhält man mit Rücksicht auf die freie Wählbarkeit der $\Delta x_\sigma$ die Identität

$$\frac{\partial^2}{\partial x_\nu \partial x_\alpha}\left(\frac{\partial \mathfrak{H}}{\partial g_\alpha^{\mu\sigma}}g^{\mu\nu}\right) \equiv 0 \cdots (90)$$

Die Gleichungen (89) und (90) sind ein Ausdruck für die Invarianz-Eigenschaften der Hamilton'schen Funktion $\mathfrak{H}$.

## Scientific creativity in the midst of personal hardships and national disaster

On the last page of the designated appendix, Einstein completes, with eq. (90), the proof of the invariance of the "action" and hence the general covariance of the field equations derived from it by the variational method. The manuscript ends somewhat abruptly without drawing the most important physical conclusion from this result. In the October paper, Einstein performs this additional step. He derives from the field equation (78) the conservation law for the combined energy-momentum of the gravitational field, emphasizing that it is derived from the field equations of gravitation alone, without using the field equations for material processes.

The year 1916 was still marked by Einstein's deliberations, publications, and correspondence with colleagues about general relativity and its consequences. Yet, in the summer of that year he published two articles that constituted an innovative and seminal contribution to the quantum theory of the interaction between electromagnetic radiation and matter. He established such basic principles as the following: (a) the absorption of radiation by an atom is proportional to the density of radiation; (b) atoms emit radiation in a spontaneous random process or in a process induced by the radiation field around them, again, with a probability proportional to the density of the radiation field; (c) in the emission and absorption process, atoms exchange with the radiation field not only energy but also momentum; (d) the emission of radiation from an atom does not spread out as a radial wave but propagates in a well-defined direction. The latter conclusion is a major confirmation of the particle (photon) nature of radiation. In September, Einstein wrote to Besso: "Thus light quanta are as good as established."

Our annotations to the manuscript pages contain background material related to relevant scientific developments and correspondence with peers. We did not address the social and political environment, nor did we talk about the deteriorating family relations during those years. How these developments affected Einstein is discussed in a number of biographies. While we have stressed that the year 1916 was a year of new beginnings and outstanding scientific creativity, it should also be mentioned that all this was achieved at a time when the Great War shook Europe and affected everyone's lives in Germany. Einstein, in particular, felt isolated because, in contrast to most of his German colleagues, he was openly critical of the war. Also, in 1916 Einstein was living alone after the breakup of his family and separation from his wife Mileva, who returned with the children to Zurich.

# NOTES ON THE ANNOTATION PAGES

MANY QUOTATIONS IN THESE PAGES ARE TAKEN FROM LETTERS AND DOCUMENTS printed in the English version of *The Collected Papers of Albert Einstein* (CPAE). Princeton, NJ: Princeton University Press.

Documents in the Albert Einstein Archives not yet printed in the CPAE appear as AEA with the archival call number.

A comprehensive four-volume work on Albert Einstein's theory of general relativity is *The Genesis of General Relativity*, ed. Jürgen Renn (Dordrecht: Springer, 2007). A number of papers from this collection are cited here.

P. 1

The English translation of Einstein's "On the Special and the General Theory of Relativity (A Popular Account)" is reprinted in CPAE vol. 6, Doc. 42, pp. 247–420. The quotation may be found on p. 312 (p. 69 of the document).

P. 2

Einstein's obituary on Ernst Mach was originally published in *Physikalische Zeitschrift* 17 (1916): 101–104. It is reprinted in CPAE vol. 6, Doc. 29, pp. 141–145.

On the epistemological deficit of classical mechanics mentioned by Einstein, see Jon Dorling, "Did Einstein Need General Relativity to Solve the Problem of Absolute Space? Or Had the Problem Already Been Solved by Special Relativity?" *British Journal for the Philosophy of Science* 29 (1978): 311–323.

P. 3

Einstein mentions the influence of Mach and Hume on his thinking in

Mach's obituary (p. 143 in reference for P. 2)
and in Albert Einstein, *Autobiographical Notes*, ed. P. A. Schilpp (La Salle, IL: Open Court [1949] 1979). The quotation appears on p. 51.

P. 4

On 14 December 1922, Einstein gave a lecture at a student reception at the Kyoto Imperial University. In this lecture, Einstein described his recollections about the genesis of the special theory of relativity and about the transition to the general theory. The notes of this lecture, "How I Created the Theory of Relativity," by Jun Ishiwara (Einstein's translator in Japan), are printed in CPAE vol. 13, Doc. 399. The quotation is on p. 638.

See also Einstein's essay "Fundamental ideas and methods of the theory of relativity, presented in their development," written December 1919/January 1920 (CPAE, vol. 7, Doc. 31, p. 21).

### P. 6

The quotation may be found on p. 235 of "Geometry and Experience" in CPAE vol. 7, Doc. 52, pp. 208–222. This is an extension of a lecture held at the Prussian Academy of Sciences, 27 January 1921.

### P. 7

The hole argument is explained in chapter 2 of this text (p. 25). See also Michel Janssen, "'No Success Like Failure . . .': Einstein's Quest for Relativity, 1907–1920," in *The Cambridge Companion to Einstein*, ed. Michel Janssen and Christoph Lehner (Cambridge: Cambridge University Press, 2014), 167–227.

### P. 9

Einstein's last paper in Prague: "On the Theory of the Static Gravitational Field," *Annalen der Physik* 38 (1912): 443–458; reprinted in CPAE vol. 4, Doc. 4, pp. 107–120.

### P. 10

This frequently quoted appeal to Grossman was transmitted orally. It appears in the reminiscences of Louis Kollros, a professor of mathematics at ETH at that time. "Erinnerungen-Souvenirs," *Schweizerische Hochschulzeitung* 28 (1955): 169–173.

Letter to Arnold Sommerfeld, 29 October 1912, in CPAE vol. 5, Doc. 421, p. 505.

See Karin Reich, *Die Entwicklung des Tensorkalküls: vom absoluten Differentialkalkül zur Relativitätstheorie* (Basel: Birkhäuser, 1994).

### P. 11

Letter to Max Laue, 22 December 1911, in CPAE vol. 5, Doc. 333, 244–245.

Einstein's dissertation is reprinted in John Stachel, *Einstein's Miraculous Year* (Princeton, NJ: Princeton University Press, 2005), 29–43.

Letter to Michele Besso, 26 March 1912, in CPAE vol. 5, Doc. 377, pp. 276–279

### P. 14

The interplay between mathematical and physical strategies is described in detail in the introduction; see also Michel Janssen and Jürgen Renn, "Untying the Knot: How Einstein Found His Way Back to Field Equations Discarded in the Zurich Notebook," in *The Genesis of General Relativity*, vol. 2, 839–925.

### P. 15

See Karin Reich, *Die Entwicklung des Tensorkalküls: vom absoluten Differentialkalkül zur Relativitätstheorie* (Basel: Birkhäuser, 1994) and Michael J.Crowe, *A History of Vector Analysis: The Evolution of the Idea of a Vectorial System* (New York: Dover, 1985).

P. 17

For further discussion on the difference between "coordinate conditions" and "coordinate restrictions," see Michel Janssen and Jürgen Renn, "Untying the Knot," pp. 839–925 (see the note for P. 14).

P. 18

The conversation between Albert Einstein and his son Eduard referred to at the end of the page was told by Einstein to journalists and also mentioned during his visit to Japan. It was quoted in many newspaper articles and books.

P. 19

The four November papers are referred to on page 105 [34]. See the note for P. 34.

P. 20

"The Formal Foundation of the General Theory of Relativity" was first published in *Königlich Preußische Akademie der Wissenschaften* (Berlin). *Sitzunsberichte* (1914): 1030–1085.

The English translation is printed in CPAE vol. 6, Doc. 9, pp. 30–83. The quotation may be found on p. 46.

For further reading on the history of parallel transport and affine connection, see John Stachel, "The Story of Newstein or: Is Gravity Just Another Pretty Force?" in *The Genesis of General Relativity*, vol. 4, pp. 1041–1078.

P. 22

In May 1921, Einstein delivered four lectures at Princeton on the theory of relativity, which have been republished many times and translated into a number of languages as *The Meaning of Relativity*. They are reprinted in CPAE vol. 7, Doc. 71. The discussion on this page refers to pp. 330–331.

Einstein in the introduction to Mario Pantaleo, *Cinquant'Anni di Relatività* (Florence: Editrice universitaria, 1955). English translation by John Stachel.

P. 23

Einstein and Grossmann's *Entwurf* theory paper is reprinted in CPAE, vol. 1, Doc. 13.

The lack of general covariance of the *Entwurf* equations is mentioned in two consecutive letters to Hendrik Lorentz:

14 August 1913, CPAE vol. 5, Doc. 467, pp. 349–351;
16 August 1913, CPAE vol. 5, Doc. 470, pp. 352–353.

Einstein's full satisfaction with the *Entwurf* theory is expressed in a letter to Michele Besso from ca. 10 March 1914, CPAE vol. 5, Doc. 514, pp. 381–382.

P. 28

For the four 1915 communications, see the note for P. 34.

Letter from Barbara Lee in *Dear Professor Einstein: Albert Einstein's Letters to and from Children*, ed. Alice Calaprice (Amherst, NY: Prometheus Books, 2002), 139–140.

P. 30

Einstein to Paul Ehrenfest, 24 January 1916 (or later), CPAE vol. 8, Doc. 185, pp. 249–254.

Einstein's correspondence with Lorentz is referred to in A. Einstein to Lorentz, 17 January 1916, CPAE, vol. 8, Doc. 183, pp. 179–181, which opens with "I am in possession of your three letters and very happy about your concurrence, . . ."

For further reading on the Lagrangian formalism, see Cornelius Lanczos, *The Variational Principles of Mechanics* (London: Dover, 1986).

P. 31

See Jürgen Renn and John Stachel, "Hilbert's Foundation of Physics: From a Theory of Everything to a Constituent of General Relativity" in *The Genesis of General Relativity*, vol. 4, pp. 857–973.

P. 32

*Autobiographical Notes*, p. 28–29 (see the note for P. 3).

P. 33

In Albert Einstein, "Physics and Reality (1936)" in *Out of My Later Years* (New York: Philosophical Library, 1950).

P. 34

The four "November papers" are reprinted in CPAE, vol. 6:

4 November 1915, "On the General Theory of Relativity," Doc. 21, pp. 98–106;

11 November 1915, "On the General Theory of Relativity (Addendum)," Doc. 22, pp. 108–110;

18 November 1915, "Explanation of the Perihelion Motion of Mercury from General Theory of Relativity," Doc. 24, pp. 112–116;

25 November 1915, "The Field Equations of Gravitation," Doc. 25, pp. 117–120.

Einstein to Hans Albert Einstein, 4 November 1915, CPAE vol. 8, Doc. 134, p. 140.

P. 35

Einstein's obituary "The Late Emmy Noether" was published as an open letter to the editor of the *New York Times*, 3 May 1935.

Einstein to Heinrich Zangger, 26 November 1915, CPAE vol. 8, Doc. 152, pp. 150–151.

On 20 November, David Hilbert presented a paper on a unified theory of electromagnetism and gravitation to the Royal Society in Göttingen. It was published on 31 March 1916 as "Die Grundlagen der Physik (Erste Mitteilung)" *Königliche Gesellschaft der Wissenschaften zu Gottingen. Mathematisch-Physikalische Klasse. Nachrichten* (1915): 395–407.

P. 36

See the "The Formal Foundation of the General Theory of Relativity" (see the note for P. 20).

Further reading: Olivier Darrigol, *World of Flow: A History of Hydrodynamics from the Bernoullis to Prandtl* (New York: Oxford University Press, 2005).

P. 37

A. Einstein, "A New Formal Interpretation of Maxwell's Field Equations of Electrodynamics," presented at the plenary session of the Royal Prussian Academy, 3 February 1916; reprinted in CPAE, vol. 6, Doc. 27, pp.132–136.

Einstein to H. A. Lorentz, 23 September 1915, CPAE vol. 8, Doc. 122, pp. 131–132.

P. 38

On 27 October 1920, at the request of Hendrik Lorentz, Einstein delivered an inaugural lecture at the University of Leiden, where he had been appointed extraordinary professor. The text of the lecture on "Ether and the Theory of Relativity" is reprinted in CPAE vol. 7, Doc. 38, pp. 161–182. The quotation is on p. 181.

Reference to the quotation from *Autobiographical Notes*, p. 33 (see the note for P. 3).

P. 39

Max von Laue, *Die Relativitätstheorie. Band 1: Die spezielle Relativitätstheorie* (Braunschweig: Friedr. Vieweg & Sohn, 1911).

For historical discussion, see Michel Janssen and Matthew Mecklenburg, "From Classical to Relativistic Mechanics: Electromagnetic Models of the Electron," in *Interactions: Mathematics, Physics and Philosophy, 1860–1930*, ed. V. F. Hendricks et al. (Berlin: Springer, 2007), 65–134.

Max von Laue to Moritz Schlick, Eggishorn, 19 August 1913 (Inv.-Nr. 108/Lau-15), Noord-Hollands Archief Haarlem (NL).

P. 40A

"Hamilton's Principle and the General Theory of Relativity." It is reprinted in CPAE vol. 6, Doc. 41, pp. 240–245.

P. 41

Einstein to Michele Besso, 21 December 1915, in CPAE vol. 8, Doc. 168, p. 163.

Karl Schwarzschild to Einstein, 22 December 1915, in CPAE vol. 8, Doc. 169, pp. 163–164.

Einstein to Schwarzschild, 9 January 1916, in CPAE vol. 8, Doc. 181, pp. 175–177.

Einstein delivered a memorial lecture on Karl Schwarzschild at an open meeting of the Royal Prussian Academy on 29 June 1916. The text is printed in CPAE (German) vol. 6, Doc. 33, pp. 358–361. It is not reprinted in the English version of CPAE.

P. 42

Einstein to George Hale, 14 October 1913, in CPAE vol. 5, Doc. 477, pp. 356–357.

Einstein to Paul Ehrenfest, 2 April 1914, in CPAE vol. 8, Doc. 2, pp. 9–10.

Einstein to Karl Schwarzschild, 9 January 1916, in CPAE vol. 8, Doc. 181, pp. 175–177.

Further reading: Klaus Hentschel, *The Einstein Tower: An Intertexture of Dynamic Construction, Relativity Theory, and Astronomy* (Palo Alto, CA: Stanford University Press, 1997).

P. 43

On 23 September 1913, Einstein delivered a lecture at the 85th meeting of the German Society of Natural Scientist and Doctors in Vienna. The lecture was published in

*Physikalische Zeitschrift* 14 (1913): 1249–62. It is reprinted in CPAE vol. 4, Doc. 17, pp. 198–222. The quoted reference to Nordström's theory is on p. 207.

P. 44

Others would argue that Einstein first rose to celebrity status after his first visit to America in 1921. See Marshall Missner, "Why Einstein Became Famous in America," *Social Studies of Science* 15: 2 (May 1985): 267–291.

For further reading, see Jean Eisenstaedt, *The Curious History of Relativity: How Einstein's Theory Was Lost and Found Again* (Princeton, NJ: Princeton University Press, 2006).

P. 45

On Mercury's perihelion rotation, see John Earman and Michel Janssen, "Einstein's Explanation of the Motion of Mercury's Perihelion," in *The Attraction of Gravitation,* ed. J. Earman, M. Janssen and J. D. Norton, vol. 5 of *Einstein Studies* (Boston: Birkhäuser, 1993) 129–172, 130, and 164n6.

P. 46

Einstein to Conrad Habicht, 24 December 1907, in CPAE vol. 2, Doc. 69, p. 47.

P. A1

"Hamilton's Principle and the General Theory of Relativity." It is reprinted in CPAE vol. 6, Doc. 41, pp. 240–245.

P. A2

Einstein to H. A. Lorentz, 19 January 1916, in CPAE vol. 8, Doc. 184, pp. 181–182.

Einstein to Hermann Weyl, 23 November 1916, in CPAE vol. 8, Doc. 278, pp. 265–266.

P. A3

Einstein to H. A. Lorentz, 13 November 1916, in CPAE vol. 8, Doc. 276, pp. 263–264.

Einstein to Michele Besso, 21 December 1915, in CPAE vol. 8, Doc. 168, p. 163.

Albert Einstein, "Über Gravitationswellen," *Königlich Preußische Akademie der Wissenschaften* (Berlin). *Sitzungsberichte* (1918): 154–167; reprinted in CPAE vol. 7, Doc. 1, pp. 9–27.

Albert Einstein and Nathan Rosen, "On Gravitational Waves," *Journal of the Franklin Institute* 223 (1937): 43–54.

Further reading: Daniel Kennefick, *Traveling at the Speed of Thought: Einstein and the Quest for Gravitational Waves* (Princeton, NJ: Princeton University Press, 2007).

P. A4

Einstein to Michele Besso, 3 January 1916, in CPAE vol. 8, Doc. 178, p. 171.

*The Special and the General Theory of Relativity: A Popular Version,* vol. 6, Doc. 42, pp. 247–417.

P. A5
Papers on the quantum theory of radiation:

"Emission and Absorption of Radiation in Quantum Theory," *Deutsche Physikalische Gesellschaft, Verhandlungen* 18 (1916): 318–323; reprinted in CPAE vol. 6, Doc 34.
"On the Quantum Theory of Radiation," *Physikalische Gesellschaft Zurich, Mittelungen* 18 (1917): 47–62; also in *Physikalische Zeitschrift* 18 (1917): 121–128; reprinted in CPAE vol. 6, Doc. 38.

Einstein to Michele Besso, 6 September 1916, in CPAE vol. 8, Doc. 254, p. 246.

# POSTSCRIPT:
# THE DRAMA CONTINUES . . .

## THE HAPPY END?

THE PUBLICATION OF EINSTEIN'S 1916 MANUSCRIPT BROUGHT AN END TO A CONVO-luted and dramatic intellectual journey that had begun almost a decade earlier. No doubt, it constituted a happy ending, because it fulfilled the highest hopes Einstein had invested into this enterprise. Indeed he had achieved a generally covariant field theory of gravitation that was both mathematically elegant and physically plausible. It also seemed to satisfy his highest philosophical ambitions, complying with the heuristic reasoning that had formed his starting point and that had been inspired by Ernst Mach ever since Einstein read his works as a student. There was no room left for Newton's metaphysical concept of absolute space in the new theory, and all its alleged physical effects, such as inertial forces, could apparently be traced back to the effect of matter.

In 1916, however, when Einstein put the final touches to his masterpiece, one could also defend a more sober perspective on what he had achieved. It hardly mattered to the world around him, busy as it was destroying European civilization in an ever more cruel and reckless war. Not even his famous academic colleagues in Berlin paid much attention to the novel theory and its implications. Einstein found it hard to attract the attention of astronomers to the obvious observational consequences of his theory such as the bending of light in a gravitational field. And gradually he himself became painfully aware that what he had elaborated with so much effort was actually different from what he had initially intended to accomplish. The tensions between Einstein's original heuristics and the implications of the new theory became ever more obvious in its further evolution.

From the drawing board to a new glimpse at the universe

## EINSTEIN'S HEURISTICS REVISITED

At first, Einstein tried tenaciously to interpret the new theory in the sense of Mach's critique of classical mechanics. There were several reasons for this tenacity, as is indicated in our commentaries. First, Mach's analysis of classical mechanics had served as an important heuristic guidance for the formulation of the theory. He had suggested that centrifugal forces might be due to an effect of the distant stars rather than Newton's absolute space. Einstein therefore naturally expected that the final theory would comply with Mach's heuristics. But he also used Mach's critique to underscore the plausibility of his audacious generalization of the classical principle of relativity to accelerated motions. In fact, there was hardly any other justification for this unconventional step. Before the spectacular confirmation in 1919 of the deflection of light in a gravitational field by a British solar eclipse expedition under the direction of Sir Arthur Eddington, such epistemological arguments played an important role. They served to highlight the advantages of the theory over competing theories that could be formulated within the framework of special relativity. Einstein's claim that the new theory agreed with the philosophical considerations of Ernst Mach thus served as a substitute for the initial lack of observational evidence.

Einstein's back against the wall—are Mach's distant stars to blame?

## A SHIFT OF ATTENTION: FROM THE GRAVITATIONAL
## FIELD EQUATION TO ITS SOLUTIONS

General relativity is a complex, mathematically sophisticated theory. Writing down the gravitational field equation does not exhaust its physical content by far, and Einstein's theory continues to surprise us to this very day. Its further mathematical elaboration, in particular, the search for exact solutions, revealed unexpected features. This process began right after its completion. Investigating mathematical aspects of the new theory in light of his original heuristics, Einstein believed, for instance, that the gravitational field is not determined uniquely by the matter distribution that serves as the source of the field equation. He concluded that the behavior of solutions at the boundaries of the universe must also be specified. But how could this demand be interpreted physically? How could it be related to the claim motivated by Mach's ideas that the inertial properties of the bodies in space are determined exclusively by the matter distribution?

After Einstein failed to find a satisfactory solution to the question of boundary conditions in regard to the Machian heuristics, in 1917 he proposed an entirely new way to solve this problem, on which he commented in a contemporary letter to his friend Ehrenfest that there was a risk it would send him to the madhouse.[1]

In Einstein's famous "Cosmological Considerations in the General Theory of Relativity" of 1917,[2] he derived a solution satisfying all his expectations concerning the constitution of the universe, including the explanation of its inertial properties by the distribution of masses acting as sources of the gravitational field. This spacetime describes a spatially closed static universe with a uniform matter distribution. With this solution Einstein avoided entirely the problem of specifying appropriate boundary conditions, since a closed space does not have a boundary. He also believed that this model corresponded to a more or less realistic picture of the universe as known at that time. It was indeed not even clear then that the galaxies observed outside our own Milky Way constituted objects of the same kind and that the universe actually extended far beyond it.

## THE EMERGENCE OF MODERN COSMOLOGY

However, these results worked for Einstein only at the price of modifying the field equation for which this static spacetime was a solution. This modification consisted of introducing an additional term, characterized by the *cosmological constant* λ, in the field equation of 1915. For Einstein's Machian philosophical conception of a static universe, the cosmological constant was a lifesaver. Eventually, however, he was forced to realize that this constant did not accomplish the purpose for which it had been invented and he abandoned it. But the additional term is really quite legitimate and contrary to what Einstein wrote in 1916, the resulting field equation was actually not the most general equation consistent with his demands. Today the cosmological constant plays an important role in explaining the accelerating expansion of the universe on the basis of the theory of general relativity.[3]

One of the most popular stories and myths about Einstein is that he referred to his idea of the "cosmological constant" as the biggest mistake he had made in his life. The origin of this story can be traced to an article in *Scientific American* (1956) by George Gamow, where the author recalled that, many years ago, he had heard Einstein's admission about this idea being "the biggest blunder he had made in his entire life." Gamow repeated this claim in his autobiography, *My World Lines*. In his recent book *Brilliant Blunders,* Mario Livio, the astrophysicist and author of popular science books, reports that there is no evidence that Einstein actually made such a statement, either orally or in writing, and that it was most probably Gamow's invention. Yet, this story has been quoted extensively; it appears in numerous books and articles and has become a generally accepted part of the Einstein scientific saga.[4]

At first Einstein tended to neglect the relation between the new theory and astronomy, at least as far as astronomy beyond the solar system was concerned. His principal opponent in the discussion of the allegedly Machian features of general relativity was the Dutch astronomer Willem de Sitter, who contributed much to making the theory known outside Germany.[5] In contrast with Einstein, de Sitter worked intensively on the astronomical consequences of the various solutions to the field equation. In an almost apologetic tone, Einstein wrote to de Sitter in 1917: "From the standpoint of astronomy, of course, I have erected but a lofty castle in the air. For me, though, it was a burning question whether the relativity concept can be followed through to the finish or whether it leads to contradictions. I am satisfied now that I was able to think the idea through to completion without encountering contradictions. Now I am no longer plagued with the problem, while previously it gave me no peace."[6]

Shortly after Einstein's "cosmological considerations" were published, de Sitter demonstrated that even the modified field equation allows a solution in which there is no matter acting as a source of the gravitational field.[7] Nevertheless, test particles moving in this spacetime do have inertial properties that cannot be explained as an effect of Mach's "distant stars." In contrast with Einstein's solution, there is no relation between the density of matter and the radius of the universe in de Sitter's solution. Both Einstein's and de Sitter's cosmological solutions became the subject of intense debate and constituted the principal alternatives that were considered. They even motivated the astronomer Edwin Hubble's observations of distant galaxies in the late 1920s, which eventually overturned the belief in a static universe, apparently against his own expectations.

Einstein's expectation of a Machian explanation of inertia in the theory of general relativity gradually changed from a requirement imposed on the theory itself to a criterion to be applied just to special solutions of the theory. He soon realized that it cannot be generally correct that inertial effects can be explained exclusively by the presence of matter in this theory. To make his heuristic expectations precise, in 1918 Einstein explicitly introduced what he called "Mach's principle."[8] It demanded that for solutions satisfying this principle, the gravitational field be completely determined by masses of bodies occurring on the right side of the field equations in the form of the energy-momentum tensor as the sources of the field. In this way Einstein translated Mach's original ideas from the language of mechanics to that of field theory.

Subsequently, Einstein began to elaborate the field theoretical interpretation of the theory of general relativity more and more—at the expense of emphasis on the mechanical roots of his original heuristics. His attitude toward Mach's ideas changed correspondingly. After 1920, the program of interpreting the theory of general relativity along the lines of Mach's philosophical critique of classical mechanics ceased to play a significant role in Einstein's research. This shift of interest was caused not least by the reorientation of his research program in the direction of a unified field theory of gravitation and electromagnetism, which had begun in 1919. In the course of his work on unified field theory, Einstein had come a long way from his initial Machian conviction that matter would play a primary role and the concept of space a derived one.

Nevertheless, the question of Mach's principle remained open, since it was now closely associated with Einstein's cosmological ideas. These largely coincided with the thinking of his contemporaries. In fact, in the period between 1917 and 1930, the prevailing topic of debate was, which static universe represents a better model of reality? At the time, the question of an expanding universe raised by Alexander Friedmann[9] in 1922 and by Georges Henri Lemaître[10] in 1927 remained largely outside the horizon of observational cosmology. Eventually, however, the ground was prepared for a decision about Mach's principle on the basis of astronomical observations.

Einstein is surprised: his "static" universe is expanding after all!

This verdict came with the accumulation of astronomical evidence in favor of an expanding universe, the decisive contribution being the work published in 1929 by Hubble,[11] who was working at the Mount Wilson Observatory. Einstein learned about these results early in 1931 during a stay at the California Institute of Technology. Almost immediately after his return to Berlin, Einstein published a paper on the cosmological problem in which he stated that Hubble's results had made his assumption of a static universe untenable.[12] Instead, as he pointed out, these results were easily explained by the dynamical solutions of the original field equation. They therefore also—at least temporarily—sealed the fate of the cosmological constant as well as that of Mach's principle. In 1954 in a letter to Felix Pirani, Einstein wrote: "In my view one should no longer speak of Mach's principle at all. It dates back to the time in which one thought that the "ponderable bodies" are the only physically real entities and that all elements of the theory which are not completely determined by them should be avoided. (I am well aware of the fact that I myself was long influenced by this idée fixe.)"[13] In an appendix to the 1954 edition of Einstein's *Relativity: The Special and General Theories* (*A Popular Account*), he returned to the issues of relativity and space, expounding his final views in a way that makes them accessible also to nontechnical readers.

## FROM THE PUZZLES OF THE SCHWARZSCHILD SOLUTION TO BLACK HOLES

What we have illustrated here for the case of Machian heuristics is equally valid for other heuristic elements that accompanied Einstein on his way to the theory of general relativity. They also turned out to require reinterpretation and revision, a process marking the ongoing conceptual development of general relativity to this day. Another famous example is the gravitational field of a single center such as a star in general relativity. As early as the winter of 1915/16, Karl Schwarzschild had derived one of the very few exact solutions of general relativity, describing the case that Einstein had previously treated by means of approximations.[14]

While Schwarzschild's work on exact solutions provided the theoretical basis for two of the three classical tests of general relativity—the perihelion motion of Mercury and the

Do not go beyond this limit! A black hole is lurking behind it.

bending of light—some aspects of its physical interpretation remained controversial for more than half a century after its formulation. In particular, the physical meaning of the so-called Schwarzschild radius was unclear. Initially, it appeared to constitute a singularity of the solution occurring when the mass of an object is concentrated in a sphere of that radius. When first confronted with this problem in 1922, Einstein was convinced that this radius simply represented a mathematical artifact, because this limit could never be reached physically.[15] In 1939, he even published a calculation by which he meant to prove that nature would not allow such strange physical behavior.[16] It was only through the concerted efforts of numerous physicists, mathematicians, and astronomers in subsequent years that the exploration of the Schwarzschild solution became connected with a thorough understanding of stellar collapse. This work eventually gave rise to the realization that our universe is not only expanding but is also filled with such apocalyptic objects as black holes.

## GENERAL RELATIVITY: FROM THE LOW WATER MARK TO THE HIGH TIDE

Immediately after its introduction, the theory of general relativity became a hallmark of international cooperation in the world of science at a time when the Great War was dividing it. The confirmation of gravitational light bending by Eddington's expedition was a culmination of this cooperative spirit. Later, the efforts devoted to the theory dwindled, partly because of the excitement generated by the rise of quantum theory. The Second World War directed the attention of physicists further away from the seemingly esoteric pursuit of the implications of Einstein's theory. The war also divided the world of science even more radically, destroying the careers and lives of many who might have otherwise collaborated in the further development of Einstein's theory. In a sense, general relativity faded into the background, being largely considered irrelevant to mainstream physics and limited to the explanation of some minor adjustments of Newton's otherwise well-confirmed theory of gravitation. The historian of science Jean Eisenstaedt has aptly characterized this period as the "Low Water Mark of General Relativity."[17] Of course, here and there some scientists achieved important insights, but these often quickly fell into oblivion. A true renaissance of general relativity began only after the war, soon to be reinforced by new astronomical discoveries. At the eve of this renaissance, John Synge described the somewhat esoteric position of general relativity in the preface of his 1960 textbook:

> Of all physicists, the general relativist has the least social commitment. He is the great specialist in gravitational theory, and gravitation is socially significant, but he is not consulted in the building of a tower, a bridge, a ship, or an airplane, and even the astronauts can do without him until they start wondering which ether their signals travel in. Splitting hairs in an ivory tower is not to everyone's taste, and no doubt many a relativist looks forward to the day when governments will seek his opinion on important questions. But what does "important" mean? Science has a dual aim: to understand nature and to conquer nature, but in the intellectual life of man surely it is the understanding which is the more important. Then let the relativist rejoice in the ivory tower where he has peace to seek understanding of Einstein's theory as long as the busy world is satisfied to do its jobs without him.[18]

All Einstein needed to reach out to the universe was chalk and a blackboard.

Soon this idyllic situation described by Synge was to change. Today, general relativity has become part of daily life: the globalized GPS technology of satellite-based navigation would not work without taking into account the implications of both special and general relativity. The tide definitely changed in the 1960s with the discoveries of quasars and of microwave background radiation, which could not have been understood without Einstein's theory. Suddenly, general relativity was back at the center stage of physics. Its apparently far-fetched mathematical constructions became much-needed tools for explaining a universe that turned out to be far more dynamic, interesting, and diverse than the static world of Einstein would have suggested.

Even today, the mathematical elaboration of the theory, the exploration of its astronomical consequences, and its physical interpretation continue to pose new problems and to generate unexpected insights. The gravitational waves predicted by the theory have been traced by astronomical observations but have not yet been confirmed by direct measurements. They will open up a new window into the universe. The efforts to achieve such measurements have definitively turned general relativity into a modern "big science," producing a flood of new results. Meanwhile, the relation of general relativity to its great sister theory in modern physics—quantum physics—has become a central concern of theoretical physicists worldwide. Nevertheless, it remains as much of a challenge now as it did when Einstein contributed to both the relativity and the quantum revolutions.

## NOTES

1. Einstein to Paul Ehrenfest, 4 February 1917, CPAE vol. 8, Doc. 294, p. 282.

2. A. Einstein, "Kosmologische Betrachtungen zur allgemeinen Relativitätstheorie," *Königlich Preußische Akademie der Wissenschaften* (Berlin). *Sitzungsberichte* (1917): 142–152; reprinted in CPAE vol. 6, Doc. 43, pp. 421–432.

3. See Malcolm Longair, *The Cosmic Century: A History of Astrophysics and Cosmology* (Cambridge: Cambridge University Press, 2006).

4. George Gamow, "Gravity," *Scientific American*, March 1961; *My World Line: An Informal Autobiography* (New York: Viking Press, 1970); Mario Livio, *Brilliant Blunders: From Darwin to Einstein; Colossal Mistakes by Great Scientists that Changed our Understanding of Life and the Universe* (New York: Simon & Schuster, 2013).

5. Janssen, Michel, "'No Success Like Failure . . .': Einstein's Quest for General Relativity, 1907–1920," in *The Cambridge Companion to Einstein*, ed. M. Janssen and C. Lehner (Cambridge: Cambridge University Press, 2014), pp. 167–227.

6. Einstein to Willem de Sitter, before March 1917, CPAE vol. 8, Doc. 311, p. 301–302.

7. See note 5.

8. A. Einstein, "On the Foundations of the General Theory of Relativity" (1918), in CPAE vol. 7, Doc. 4, pp. 33–35.

9. Alexander Friedmann, "Über die Krümmung des Raumes," *Zeitschrift für Physik* 10:1 (1922): 377–386.

10. Georges Lemaître, "Un Univers homogène de masse constante et de rayon croissant rendant compte de la vitesse radiale des nébuleuses extra-galactiques," *Annales de la Société Scientifique de Bruxelles* 47 (April 1927): 49.

11. Hubble, Edwin, "A Relation between Distance and Radial Velocity among Extra-Galactic Nebulae," *Proceedings of the National Academy of Sciences* 15:3 (1929): 168–173.

12. A. Einstein, "Zum kosmologischen Problem der allgemeinen Reitivitätstheorie," *Sitzungsberichte der Preußische Akademie der Wissenschaften* (1931): 235–237. See also Harry Nussbaumer, "Einstein's Conversion from His Static to an Expanding Universe," *European Physical Journal H* 39 (2014): 37–62.

13. Einstein to Felix Pirani, 2 February 1954 (AEA 17—447.00).

14. Karl Schwarzschild, "Über das Gravitationsfeld eines Massenpunktes nach der Einsteinschen Theorie," *Sitzungsberichte der Königlich Preußischen Akademie der Wissenschaften* 1916: 189–196.

15. Jean Eisenstaedt, "The Early Interpretation of the Schwarzschild Solution," in *Einstein and the History of General Relativity: Based on the proceedings of the 1986 Osgood Hill Conference, North Andover, Massachusetts, 8–11 May 1986*, vol. 1 of *Einstein Studies* (Basel: Birkhäuser, 1986), 213–233

16. A. Einstein, "On a Stationary System with Spherical Symmetry of Many Gravitating Masses," *Annals of Physics* 40 (1939): 922–936.

17. Jean Eisenstaedt, "The Low Water Mark of General Relativity, 1925–1955," in *Einstein and the History of General Relativity*, vol. 1 of *Einstein Studies*, 277–292.

18. John L. Synge, *Relativity: The General Theory* (Amsterdam: North-Holland, 1960).

# A CHRONOLOGY OF THE GENESIS
## OF GENERAL RELATIVITY AND
## ITS FORMATIVE YEARS

*Between 1902 and 1909* Einstein works at the Swiss Patent Office in Bern.

*1905, June 30*, Einstein submits "On the Electrodynamics of Moving Bodies," the first formulation of the special theory of relativity.

*1905, July*, Henri Poincaré proposes two laws of gravitational attraction compatible with a special relativistic framework and all astronomical observations explained by Newton's law.

*1905, September 27*, Einstein submits "Does the Inertia of a Body Depend upon Its Energy Content?" The paper introduces the notion that mass is a measure of the energy content of a body.

*1906, May 17*, Einstein submits "The Principle of Conservation of Motion of the Center of Gravity and the Inertia of Energy." The paper shows that the inertia of a body depends on its energy content.

*1907, May 14*, Einstein submits "On the Inertia of Energy Required by the Relativity Principle." For the first time he speaks of "the equivalence of mass and energy," but the implications for the relation between inertial and gravitational mass are not yet addressed.

*1907, November 5*, Hermann Minkowski discusses Poincaré's laws of gravitation and introduces his four-dimensional spacetime formalism.

*1907, December 4*, Einstein submits "On the Relativity Principle and the Conclusions Drawn from It," a review paper in which he first explores the implications of the new kinematics of special relativity for gravitation and introduces the equivalence principle, as well as its immediate observational consequences.

*1907, December 24*, Einstein writes to his friend Conrad Habicht that he is trying to explain the shift of Mercury's perihelion on the basis of a relativistic treatment of the law of gravitation but that he has so far not succeeded.

*1908, September 21*, Hermann Minkowski delivers a lecture on spacetime, discussing the compatibility between his four-dimensional framework of special relativity and Newton's law of gravitational attraction.

*1909, September*, Einstein begins to consider an extension of the relativity principle to uniformly rotating systems.

*1909, October 15*, Einstein assumes duties as extraordinary professor at the University of Zurich.

*1910*, Einstein embraces Minkowski's four-dimensional formalism as an essential springboard to the extension of the special theory of relativity.

*1911, April 1*, Einstein assumes his appointment as professor at the German University of Prague.

*1911*, Max von Laue finds that the inertial behavior of an extended physical system in special relativity has to be described by the energy momentum tensor that later will play the major role as a source of the gravitational field.

*1911, June 21*, Einstein submits "On the Influence of Gravitation on the Propagation of Light." He predicts that the gravitational bending of light rays passing near the sun may be confirmed by astronomical observations.

*1911, December 14*, Max Abraham submits the first of a series of papers in which he develops a theory of the gravitational force using Minkowski's four-dimensional spacetime formalism.

*1912, February 15*, Max Abraham reacts to Einstein's criticism that his theory is actually incompatible with Minkowski's framework by introducing an infinitesimal line element with variable metric but without further commenting on it.

*1912, February 26*, Einstein submits "The Speed of Light and the Statics of the Gravitational Field." In this paper he proposes a relativistic theory of the static gravitational field based on a generalization of Newton's gravitational law using the equivalence principle. By this time, he must also have realized that a complete theory of gravitation would need to go beyond a scalar theory and also beyond Euclidean geometry.

*1912, March 23*, Einstein submits "On the Theory of the Static Gravitational Field." He corrects his earlier theory, which he had found not to be in agreement with the principle of momentum conservation. Realizing that the gravitational field can act as its own source, he amended his earlier field equation.

*1912, April 15–22*, Einstein visits the astronomer Erwin Freundlich in Berlin and discusses with him the possibilities to observe gravitational bending, for instance, during a solar eclipse. They also discuss gravitational redshift, as well as the idea of gravitational lensing, which Einstein publishes only 24 years later.

*1912, May 23*, Einstein publishes a "Note Added in Proof" to his earlier paper on the static gravitational field, reformulating the equation of motion in a gravitational field with the help of a variational principle. At the end of the note he states that this reformulation suggests how such an equation should look for the general case, thus pointing to the role of the line element in general relativity.

*1912, July*, Einstein publishes "Is there a Gravitational Effect Which is Analogous to Electrodynamic Induction." This paper makes it clear that Mach's ideas and the analogy with electromagnetism are important guidelines for his search for a relativistic theory of the gravitational field.

*1912, July 25*, Einstein departs for Zurich to take up a position as professor at the ETH.

*1912, August*, Einstein establishes an expression for the general equation of motion in a gravitational field.

*Summer 1912 to Spring 1913*, Einstein explores with Grossmann the resources of Riemannian geometry to find a gravitational field equation. This effort is documented in the famous Zurich Notebook.

*1912, 20 October*, Gunnar Nordström publishes a new gravitation theory within a special relativistic framework.

*1913, before May 28*, Einstein and Grossmann complete "Outline of a Generalized Theory of Relativity and of a Theory of Gravitation" (the *Entwurf* theory). The theory fails to be generally covariant, but Einstein eventually convinces himself that this is unavoidable.

*1913, May*, with the help of Michele Besso, Einstein derives the perihelion shift of Mercury based on the *Entwurf* theory, which leads to about half of the correct result.

*1913, after August*, following a suggestion by Michele Besso, Einstein develops the hole argument, which seemingly excludes generally covariant theories.

*1913, September 23*, Einstein lectures in Vienna on the "On the Present State of the Problem of Gravitation." Here, he refers to Nordström's special relativistic gravitation theory as the only viable competitor.

*1913, December*, in a letter to Ernst Mach, Einstein suggests that Mach's criticism of Newton's concept of absolute space is the strongest support of his theory.

*1914, January 30*, Einstein adds the hole argument as an additional argument to justify the *Entwurf* theory.

*1914, February 19*, Einstein publishes, jointly with A. D. Fokker, "Nordström's Theory of Gravitation from the Point of View of the Absolute Differential Calculus."

*1914, March 29*, Einstein arrives in Berlin to take up his position at the Royal Prussian Academy of Sciences.

*1914, May 29*, Berlin, Einstein publishes, jointly with Grossmann, "Covariance Properties of the Field Equations of the Theory of Gravitation Based on the Generalized Theory of Relativity," in which he introduces a variational principle for the *Entwurf* theory.

*1914, October 29*, Berlin, Einstein submits "The Formal Foundation of the General Theory of Relativity" as a conclusive account of general relativity based on the *Entwurf* theory, a claim that he will soon regret.

*1915, June 28–July 5*, Göttingen, Einstein gives several lectures on general relativity.

*1915, November 4*, Einstein abandons the *Entwurf* theory and submits to the Prussian Academy of Sciences the first of a series of papers, titled "On the General Theory of

Relativity." There he returns to a gravitation theory based on the Riemann tensor, however, without yet reaching general covariance.

*1915, November 11*, Einstein submits "On the General Theory of Relativity (Addendum)." This paper offers a reinterpretation of his earlier results, introducing the hypothesis that all matter is of electromagnetic origin.

*1915, November 18*, Einstein submits "Explanation of the Perihelion Motion of Mercury from the General Theory of Relativity," in which a calculation based on the new theory provides the expected result.

*1915, November 25*, Einstein submits the definitive version of general relativity in a paper titled "The Field Equations of Gravitation."

*1915, December 22*, Karl Schwarzschild communicates to Einstein the first exact solution of general relativity that describes a spherically symmetric gravitational field in a vacuum.

*1915, December 26*, in a letter to Paul Ehrenfest Einstein outlines the first coherent synthesis of his new theory presented in the four November papers.

*1916, February 24*, Einstein submits to the academy Schwarzschild's second paper, describing the interior gravitational field of a sphere of fluid with uniform energy density. Here the quantity later to become known as the Schwarzschild radius—which will play an important role in the theory of black holes many decades later—makes its first appearance.

*1916, March 20*, Einstein submits "The Foundation of the General Theory of Relativity." An unpublished "Appendix" supplies his "Formulation of the Theory on the Basis of a Variational Principle."

*1916, June 22*, Einstein presents his "Approximate Integration of the Field Equations of Gravitation," the first paper in which he raises the possibility of gravitational waves.

*1916, October 26*, Einstein publishes a slightly modified version of the unpublished appendix to the "The Foundation of the General Theory of Relativity," titled "Hamilton's Principle and the General Theory of Relativity."

*1916, December*, Einstein finishes his exposition "On the Special and the General Theory of Relativity (A Popular Account)."

*1917, February 8*, Einstein submits the paper "Cosmological Considerations in the General Theory of Relativity," in which he introduces the cosmological constant to secure a static universe compatible with Mach's idea that inertia is caused by cosmic masses.

*1918, January 31*, Einstein submits his paper "On Gravitational Waves."

*1918, March 6*, Einstein submits "On the Foundations of the Theory of General Relativity," the paper in which he explicitly introduces Mach's principle as a criterion for admissible solutions of general relativity. He also agrees with Erich Justus Kretschmann, who argued in 1917 that every meaningful physical theory can be expressed in a generally covariant form.

*1919, September 22*, Einstein receives confirmation of the predicted bending of light by the gravitational field of the sun, observed during a solar eclipse.

*1921, April 2 to May 30*, Einstein lectures at Princeton University. His lectures are later turned into a formative text published under the title "The Meaning of Relativity."

*1922, November 9*, Einstein is informed that he is to receive the 1921 Nobel Prize in Physics.

*1922, September*, Alexander Friedmann publishes a dynamic solution of the field equation that is criticized by Einstein, who later retracts his criticism.

*1927 April*, Georges Lemaître publishes a paper on a solution of general relativity describing an expanding universe.

*1929 January*, Edwin Hubble publishes a paper on his astronomical observations suggesting an expansion of the universe.

*1929*, beginning of an exchange of letters between Einstein and Elie Cartan dealing, among other subjects, with a mathematical reformulation of general relativity in the context of attempts at its generalization.

*1931, April 16*, Einstein publishes a paper on the cosmological problem, in which he states that Hubble's results have made his assumption of a static universe untenable.

*1932, March*, in a joint paper with Willem de Sitter he withdraws the cosmological constant.

# PHYSICISTS, MATHEMATICIANS, AND PHILOSOPHERS RELEVANT TO EINSTEIN'S THINKING

## Abraham, Max (1875–1922)

The theoretical physicist Max Abraham worked predominantly on Maxwell's theory of electricity. In 1902 he developed a theory of electrodynamics in which he applied Maxwell's differential equations of the electromagnetic field to the dynamics of electrons. Abraham's theory was based on the assumption of an ether in which the electromagnetic phenomena took place, as opposed to Einstein's special theory of relativity published in 1905. In the following years the two were involved in a fundamental scientific dispute. Abraham mastered the mathematical technicalities and was able to fully understand Einstein's theory, but he refused to accept it on the basis of his physical assumptions about the nature of the ether and the electron. Abraham was the first to use Minkowski's formalism to propose a gravitational field theory, which provoked another controversy with Einstein.

Photo: Niedersächsische Staats- und Universitäts- bibliothek Göttingen

## Bernays, Paul (1888–1977)

The Swiss mathematician Paul Bernays was lecturer of analysis at the University of Zurich from 1912 to 1919. Afterward, he taught at the University of Göttingen until 1933, when he was forced to emigrate and returned to Zurich. Bernays's main contributions concerned mathematical and propositional logic, axiomatic set theory, and the foundations of mathematics. In 1914, he suggested to Einstein and Grossman that they use variational calculus in the formulation of the relativity theory. During his time in Göttingen, Bernays collaborated with Hilbert in pursuing the program of axiomatization of mathematics and contributed to Hilbert's main book on the foundations of mathematics.

Photo: ETH-Bibliothek Zürich, Bildarchiv

## Besso, Michele (1873–1955)

Einstein met Michele Besso, a mechanical engineer, in 1896 in Zurich, where Besso had recently enrolled as a student at the Federal Technical University. They became lifelong friends and, for a while, were even colleagues at the Swiss Patent Office in Bern. Besso was an interested listener who was able to pose questions that would stimulate Einstein, in their frequent and long conversations, to discuss, clarify, and develop his ideas. Besso's role in this process was so important that he was the only person Einstein acknowledged in his paper introducing the special theory of relativity.

Photo: Courtesy of the Besso family, Lausanne, Schweiz

### Born, Max (1882–1970)

Max Born was one of the founders of modern quantum physics. Born studied mathematics and physics at different universities. Among his teachers in Göttingen was Hermann Minkowski, who introduced him to electrodynamics and special relativity theory. Born's first publications, written between 1909 and 1914, were dedicated to electron theory, relativity theory, crystal physics, and Einstein's quantum theory of specific heat. In the following years, he focused his work on atomic physics and the mathematical development of quantum physics. In 1915, Born was appointed professor for theoretical physics at the University of Berlin, where he became a close friend of Einstein's. Later, he also taught in Frankfurt, and from 1921 in Göttingen, where he formed a group that in 1925 formulated the foundations of quantum mechanics. In 1933, Born was forced to emigrate and moved to Great Britain.

### Christoffel, Elwin Bruno (1829–1900)

The German mathematician Elwin Christoffel contributed with several works to the development of Bernhard Riemann's theory of curved surfaces (Riemannian geometry). In particular, in 1882, he introduced index symbols that describe the transformation of geometric data along curved surfaces. In 1901, the Christoffel symbols were incorporated into the absolute differential calculus by Gregorio Ricci-Curbastro and Tullio Levi-Civita. In 1912, Marcel Grossman acquainted Einstein—at the time working on the field equations of his gravity theory—with this mathematical formalism. But it was not until 1915, when Einstein interpreted Christoffel symbols as a mathematical expression of the gravitational field, that he was able to formulate the field equations of the general theory of relativity.

### Eddington, Arthur Stanley (1882–1944)

The British astrophysicist Arthur Eddington became director of the Observatory of Cambridge in 1914. His research focused on stellar physical processes such as radiation and energy generation and, later, on the mathematical and cosmological aspects of general relativity. Eddington became aware of Einstein's general theory of relativity in 1915 through Willem de Sitter's papers and soon began to promote its empirical testing. In 1919, he led an expedition to the African island of Principe to observe the behavior of light rays in the gravitational field of the sun during a solar eclipse. Eddington's observations confirmed the deflection of light as predicted by Einstein's theory and were considered at the time conclusive proof of general relativity. In 1979, Eddington's results were confirmed using modern measuring equipment.

## Ehrenfest, Paul (1880–1933)

In 1912, Paul Ehrenfest was appointed to the Chair of Theoretical Physics at Leiden, succeeding Hendrik Antoon Lorentz. In the preceding years he and his wife Tatiana Afanasieva had made important contributions to statistical mechanics. In 1909, Ehrenfest formulated a paradox showing that rigid bodies are not compatible with the special theory of relativity. This led Einstein to recognize that the geometry of a rotating frame of reference is not Euclidean. Ehrenfest worked predominantly on the early quantum theory and later on quantum mechanics. After meeting for the first time in 1912, Einstein and Ehrenfest discussed various physics problems on many occasions. Ehrenfest also arranged a number of notable conversations between Einstein and Niels Bohr on quantum physics.

## Eötvös, Lorand (1848–1919)

The Hungarian physicist Lorand Eötvös first gained international recognition with his contributions to the study of capillarity, but gravitation and its measurement soon became the lifelong focus of his work. To measure the gravitational attraction exerted by Earth on different substances or at different locations, Eötvös employed the torsion balance, an instrument previously used by Henry Cavendish to determine the attraction between two masses. Eötvös developed a complete theory of the torsion balance, improved its sensitivity, and devised new methods of measurement. In particular, he performed a long series of very precise measurements of the rate of gravitational acceleration for different bodies that proved the equivalence of gravitational mass and inertial mass. The principle of equivalence based on it became one of the building blocks of the theory of general relativity.

Photo: © Foto Deutsches Museum

## Euler, Leonhard (1707–1783)

The Swiss mathematician Leonhard Euler was a member of the Russian Academy of Science in Saint Petersburg from 1727 to 1741 and again from 1766 onward. In the period between these years, he worked at the Prussian Academy of Science in Berlin. Euler made enormous contributions to almost all areas of mathematics, such as analysis, infinitesimal calculus, graph theory, geometry and trigonometry, differential and integral calculus, algebra, and number theory. He also introduced the modern mathematical terminology and notation. Euler's work on mathematics was closely connected with applications to problems of technology, astronomy, and physics—for example, optics, statics, and hydraulics. In particular, he created the mathematical apparatus of hydrodynamics, including the equations governing fluid motions used by Einstein in the formulation of his relativity theory.

### Freundlich, Erwin Finlay (1885–1964)

Erwin Freundlich was a young assistant at the Astronomical Observatory in Berlin when he first came into contact with Einstein in 1911 through the Prague astronomer Leo Wenzel Pollak. Freundlich enthusiastically took on the task of testing the effects of gravitation on the propagation of light, as predicted by the still-incomplete theory of general relativity. For a while, he was one of Einstein's closest collaborators but remained isolated from other German astronomers because of their skepticism toward Einstein's theory. In 1920, he was the principal initiator of the Einstein Institute, an observatory dedicated specifically to strengthening the empirical foundations of Einstein's gravitational theory in the astrophysical domain. During his scientific career, Freundlich organized several expeditions to observe light deflection during solar eclipses and focused his research on the wavelength shifts in solar and stellar spectra.

### Friedmann, Alexander Alexandrowitsch (1888–1925)

The Russian physicist Alexander Friedmann taught at the University of Perm from 1918 until 1920 and subsequently worked at the Academy of Science in Petrograd, now Saint Petersburg. He concentrated his research on theoretical meteorology and hydromechanics but also worked on mathematical and cosmological aspects of the general theory of relativity. In two papers published in 1922 and in 1924, Friedmann outlined a model of a nonstatic universe in the frame of general relativity, dispensing with Einstein's cosmological constant in the field equations. Einstein later acknowledged that Friedmann's solution was correct, though Einstein did not accept it, as he was not ready to renounce the model of a static universe.

### Gamow, George (1904–1968)

The Russian theoretical physicist George Gamow immigrated to the United States in 1934 and was appointed professor of physics at George Washington University in Washington, DC. After an initial phase dedicated to the research of problems of quantum theory and nuclear physics, Gamow concerned himself with cosmology and the application of nuclear physics to astronomical phenomena. He was a strong advocate of the expanding-universe theory and contributed to the development of the Big Bang theory regarding the origin of the universe. Gamow wrote a great number of successful popular science books.

### Gauss, Carl Friedrich (1777–1855)

For most of his life Carl Friedrich Gauss was the director of the astronomical observatory in Göttingen but also worked in a wide range of scientific domains, such as mathematics, geometry, mechanics, and dioptrics, to which he made substantial contributions. In particular, Gauss dealt also with geodesy and in 1828 published a book on the geometry of curved surfaces that became the fundament of the new discipline of differential geometry. In 1912, Einstein became aware of the analogy between the mathematical formulation of general relativity and the Gaussian theory of curved surfaces. Shortly afterward, his friend Marcel Grossman introduced him to the modern developments in differential geometry, whose new mathematical instruments enabled Einstein to finalize his theory.

### Grossmann, Marcel (1878–1936)

Einstein and Marcel Grossmann became close friends as students at the Federal Technical University in Zurich. Grossmann studied mathematics and in 1907 became professor of descriptive geometry at the same university. In the following years, he helped Einstein not only in his academic career but also in his scientific work. In 1912, Grossman introduced Einstein to recent developments in the absolute differential calculus by Riemann, Ricci-Curbastro, and Levi-Civita. Together they developed new mathematical tools and formulated a first generalized theory of relativity, the so-called *Entwurf* theory.

Photo: From *Proceedings of the First Marcel Grossmann Meeting on General Relativity*, Trieste, 1975.—NH, 1977.—ISBN 0720407079

### Hale, George Ellery (1868–1938)

At a time when astronomers were mostly concerned with the positions, motions, and distances of stars rather than their physical nature, in his student years, the U.S. astronomer George Ellery Hale was already applying spectroscopic methods to the observation of solar phenomena. Hale made great contributions to the advancement of astrophysical research with his observations, in particular on solar spectra and the magnetic fields in sunspots, and by promoting the foundation of astrophysical observatories. In 1904, he founded the Mount Wilson Observatory and directed it until his retirement. It was Hale's authority in solar astrophysics that probably led Einstein in 1913 to seek his advice on the possibility of detecting light deflection in the sun's gravitational field.

### Hertz, Heinrich Rudolf (1857–1894)

The German physicist Heinrich Hertz was the first to prove conclusively the existence of electromagnetic waves predicted by James C. Maxwell. Hertz built an experimental setup to transmit and receive radio pulses using procedures that ruled out all other possible wireless phenomena. He also developed a formulation of classical mechanics excluding the concept of force.

### Hilbert, David (1862–1943)

In 1895, David Hilbert was appointed professor of mathematics at the University of Göttingen. He was one of the most influential mathematicians of his time and contributed substantially to many branches of his discipline, including the theory of invariants, algebraic number theory, analysis, and the theory of integral equations. Around 1898–1899, Hilbert worked on the axiomatic foundations of geometry. Pursuing a program of the axiomatization of all mathematical sciences, in 1912 Hilbert started to work on the axiomatic foundations of physics and learned of Einstein's efforts toward general relativity. In 1915, he integrated Einstein's results in a unifying theory of electrodynamics and relativity.

### Hubble, Edwin (1889–1953)

Photo: Hale Observatories, courtesy AIP Emilio Segre Visual Archives

The U.S. astronomer Edwin Hubble began working in 1919 at the Mount Wilson Observatory in California. With his investigations on galactic nebulae and variable stars, he played a crucial role in establishing modern extragalactic astronomy. In a paper published in 1929, Hubble delivered observational evidence that the recessional velocity of a galaxy increases with its distance from the observer. Initially, he was uncommitted to the conclusion of an expanding universe, as expounded some years earlier by Georges Lemaître and Alexander Friedmann on the basis of calculations in the frame of Einstein's general relativity. Hubble's results convinced Einstein to abandon the cosmological constant he had introduced in his field equations to maintain the model of a static universe.

### Hume, David (1711–1776)

David Hume was a major figure in what is known as the Scottish Enlightenment, the blossoming of culture and sciences that took place in particular in Edinburgh and Glasgow in the mid-eighteenth century. Hume was a historian and, above all, a philosopher. His reflections concerned political theory, economics, ethics, logic, and theory of knowledge. According to Hume, all ideas are based on perceptions—that is, experience through senses—and knowledge is the result of experimental reasoning—that is, reflection on data of experience. It is therefore not possible to assert something about what one cannot experience. As to scientific thinking, Hume maintained that the necessary connections between facts, scientific laws or causal connections, are constructs of the human mind resulting from repeated experiences and do not have a metaphysical existence.

### Kottler, Friedrich 1886–1965

Friedrich Kottler was professor of mathematical physics at the Philosophical School of the University of Vienna. He was an important contributor to the further development of special theory of relativity. Building on the work of Minkowki, Sommerfeld, and Laue on four-dimensional electrodynamics, Kottler was the first in 1912 to use the absolute differential calculus of Ricci-Curbastro and Levi-Civita to express Maxwell's equations in generalized coordinates. However, he did not relate this work to the problem of gravitation. Einstein later used Kottler's work in developing the general relativity theory. After immigrating to the United States in 1939, Kottler worked for many years as a chemist at Eastman Kodak. He returned to Austria in 1956, where he reclaimed his chair at the university.

Photo: AIP Emilio Segre Visual Archives, Physics Today Collection

### Kretschmann, Erich Justus (1887–1973)

The German physicist Erich Kretschmann studied with Max Planck and Heinrich Rubens at the University of Berlin where he received his Ph.D. in 1914. He became a high school teacher and later a professor of theoretical physics in Königsberg and Halle. In 1917 he published a paper on Einstein's theory of general relativity in which he argued that the principle of general covariance has no physical content but just constitutes a mathematical requirement, a claim that led to an exchange with Einstein about the meaning of the general principle of relativity.

Photo: Universitätsarchiv Halle, Rep. 40/I, K 62

### Lemaître, Georges (1894–1966)

Alongside theological studies, the Belgian Catholic priest Georges Lemaître pursued research in astrophysics, cosmology, and mathematics. In 1927, he was appointed professor of physics at the University of Louvain. As early as 1925, Lemaître worked on the application of Einstein's general theory of relativity to cosmology and in 1927 published a fundamental paper in which he solved the equations of the gravitational field without using Einstein's cosmological constant. He maintained that the universe is expanding, as Alexander Friedmann had shown some years earlier. Lemaître delivered a demonstration of Hubble's law of recessional velocity before Hubble himself did. Einstein and Lemaître discussed the theory on several occasions, but Einstein accepted its cosmological consequences only in 1931, when he learned of Hubble's results.

### Levi–Civita, Tullio (1873–1941)

The Italian mathematician Tullio Levi-Civita taught at the Universities of Pavia, Padua and, finally, Rome. He published a great number of works on pure and applied mathematics, dealing in particular with analytical mechanics, celestial mechanics, hydrodynamics, elasticity, electromagnetism, and atomic physics. Around 1899–1900, Levi-Civita and his professor, Ricci-Curbastro, wrote a fundamental treatise on absolute differential calculus and its applications to express geometric and physical laws in Euclidean and non-Euclidean spaces. Einstein and Grossmann later used and developed these new mathematical tools to formulate the general theory of relativity. In the years 1915–1917, Einstein and Levi-Civita corresponded about mathematical problems of the theory.

### Lorentz, Hendrik Antoon (1853–1928)

The Dutch physicist Hendrik Antoon Lorentz was appointed professor at the University of Leiden in 1877. His most influential contributions concerned the theory of light and electromagnetism as well as the electron theory. Starting from Maxwell's electromagnetic theory, Lorentz developed his electron theory and, in particular, an electrodynamics of moving bodies based on the assumption of a stationary ether. Lorentz's electrodynamics constituted the fundament of Einstein's special theory of relativity, which, however, disposed of the medium ether. Lorentz recognized the consistency of Einstein's theory and made contributions to it but continued to support the existence of the ether. In later years, Lorentz also contributed to the development of the general theory of relativity.

### Mach, Ernst (1838–1916)

The Austrian physicist and philosopher Ernst Mach was appointed professor of physics at the University of Prague in 1867, and professor for history and philosophy of science at the University of Vienna in 1895. His physical research was dedicated to problems of optics (Doppler effect) and acoustics (sound waves). In this connection, Mach also studied the physiology and psychology of sensory perception, on which he published a book in 1886. These studies led Mach to question the mechanistic and atomistic views prevalent at the time and to formulate an empirical theory of knowledge that strongly influenced the logical positivism of the twentieth century. In particular, Mach's criticism of Newton's concept of absolute space, expounded in a book he published in 1883 on the history of mechanics, played a role in Einstein's reflections that led him to the general theory of relativity.

### Maxwell, James Clerk (1831–1879)

Besides his contributions to geometric optics, kinetic theory of gases, and thermodynamics, as well as to other fields of theoretical and experimental physics, the British physicist James Clerk Maxwell is best known for his investigations in electromagnetism. In the years 1860–1862, he developed a first formulation of a set of equations that describe electromagnetic and optic phenomena as manifestations of the electromagnetic field and suggested the hypothesis that light itself consists of electromagnetic waves. Maxwell's equations of the electromagnetic field became a starting point for Einstein's special theory of relativity.

### Mie, Gustav (1869–1957)

The German physicist Gustav Mie was appointed professor at the University of Greifswald in 1902 and later taught at the universities of Halle and Freiburg. Besides his contributions concerning optical phenomena in colloids, X-ray analyses of organic compounds, and electromagnetism, Mie's most influential work was a theory of matter published in 1912. It represents a nonlinear extension of Maxwell's electrodynamics in the framework of Einstein's special relativity, in the hope of explaining particles as emerging properties of the field. Mie also tried to include gravitation in his electrodynamics, an attempt that was later taken up by Hilbert.

Photo: Leopoldina Archiv, 02/06/64/70, MM 3412

### Minkowski, Hermann (1864–1909)

Hermann Minkowski was professor of mathematics at the Federal Technical University in Zurich from 1896 until 1902 and afterward at the University of Göttingen. During his studies in Zurich, Einstein had attended several of his courses. Minkowski created and developed the geometry of numbers, a geometric method of solving problems in number theory. In addition, he applied this approach to the field of mathematical physics and the theory of relativity. In 1907, he showed that Einstein's special theory of relativity could be understood geometrically as a theory of four-dimensional spacetime. Minkowski's geometric formulation of the theory became the basis for Einstein's formulation of generalized theory of relativity.

### Nernst, Walther (1864–1941)

Walther Nernst was appointed professor of physical chemistry at the University of Berlin in 1905 and led research institutes for physical chemistry and for applied physics. With his works on electrochemistry, solid-state chemistry, photochemistry, and the thermodynamic of gases, Nernst contributed to the establishment of physical chemistry. In 1905, he formulated the so-called third law of thermodynamics. The experiments to confirm the law also led to the confirmation of Einstein's quantum theoretical predictions on the specific heat of solid bodies. Nernst was a great organizer of scientific research and initiator of scientific institutions. As a member of the Prussian Academy of Science, he played a decisive role in Einstein's election to the academy and his subsequent move to Berlin.

### Noether, Emmy (1882–1935)

The German mathematician Emmy Noether is considered as one of the founders of modern algebra, but she had to overcome prejudice and opposition to women's rights to study and to pursue an academic career before she was able to obtain a professorship at the University of Göttingen in 1922. By that time, Noether had already made significant contributions to the theory of algebraic invariants, such as the so-called Noether's theorem, published in 1918, which provides a general mathematical demonstration also for the conservation laws in Einstein's general theory of relativity. In the following years, Noether contributed to topology and other fields of mathematics. Above all, she created and developed the new field of abstract algebra.

### Nordström, Gunnar (1881–1923)

The Finnish theoretical physicist Gunnar Nordström became a lecturer at the University of Helsinki in 1910 and later professor at the Technical University there. As a student, he had already written about Minkowski's electrodynamics and about the theory of relativity. In 1912, he formulated a special relativistic theory of gravitation and in the following years published several works concerning this topic. Nordström himself discussed this subject with Einstein on several occasions, for instance, in Zurich in 1913. Later, Nordström did research on radioactivity and also worked on electricity theory and thermodynamics.

### Ricci–Curbastro, Gregorio (1853–1925)

The Italian mathematician Gregorio Ricci-Curbastro began teaching at the University of Padua in 1880 and remained there for 40 years. He published works on higher algebra, on infinitesimal analysis, and on the theory of real numbers but is best known for the invention of the absolute differential calculus, a calculus in which formulas and results retain the same form no matter what system of variables is used. In 1900, Ricci-Curbastro published a complete exposition, written with his student Levi-Civita, about the calculus and its applications, for example, in geometry and in particular to the Riemannian manifold. In 1912, Grossmann realized that the mathematical language of absolute differential calculus could be used to formulate Einstein's general theory of relativity.

### Riemann, Bernhard (1826–1866)

The German mathematician Bernhard Riemann began to study mathematics in Göttingen under Carl Friedrich Gauss and later took over his chair. He made lasting contributions to analysis, number theory, and differential geometry. In 1854, he extended Gauss's differential geometry, which dealt with curved surfaces in a three-dimensional space, to a theory of curved surfaces in $n$-dimensional spaces known as Riemannian geometry, thereby setting the stage for Einstein's general theory of relativity. In order to characterize such curved surfaces, he introduced the curvature tensor that became an important mathematical instrument of general relativity.

### Schlick, Moritz (1882–1936)

The German philosopher Moritz Schlick was the founder of the Vienna Circle. After completing his studies in physics, Schlick turned to philosophy and concerned himself with problems of ethics, epistemology, and philosophy of science. In 1922, he was appointed professor at the University of Wien, as successor to Ernst Mach. In 1915, Schlick had already published a paper on the philosophical implications of Einstein's theory of relativity, discussing Einstein's clarification of the concept of simultaneity at a distance. In a subsequent book about the concepts of space and time in contemporary physics, Schlick explained Einstein's adoption of non-Euclidean geometry. In 1918 he published a major work on the theory of knowledge. He argued that the physically real is characterized by space-time coincidences, a concept that became instrumental in overcoming Einstein's infamous hole argument.

### Schwarzschild, Karl (1873–1916)

The German astrophysicist Karl Schwarzschild was professor of astronomy at the University of Göttingen and director of the local observatory from 1901 to 1909. Later, he became director of the astrophysical observatory in Potsdam. In his early years, Schwarzschild focused his work on celestial mechanics and stellar photometry. He was the first to systematically use photography in astronomical observations. Later, he also concerned himself with problems of geometric optics, electrodynamics, and spectroscopy applied to astrophysical phenomena. In 1914, Schwarzschild attempted to observe a gravitational redshift in the solar spectrum, as predicted by Einstein's general theory of relativity. In 1915 and 1916, he was the first to develop exact solutions to Einstein's field equation.

### de Sitter, Willem (1872–1934)

In 1908, the Dutch astronomer Willem de Sitter was appointed professor at the University of Leiden, where he became director of the observatory in 1919. His main contributions concerned celestial mechanics, stellar photometry, the measurement of stellar parallaxes, and the application of the theory of relativity to cosmology. In 1913, de Sitter provided astronomical evidence for the constancy of the velocity of light in observations of double stars, thus confirming Einstein's special theory of relativity. In the years 1916–1917, he published papers on the astronomical consequences of Einstein's gravitational theory, which attracted the attention of Arthur Stanley Eddington and led to his solar eclipse expedition in 1919. In 1932, de Sitter and Einstein collaborated on a paper about the expansion of the universe.

### Sommerfeld, Arnold (1868–1951)

The German theoretical physicist Arnold Sommerfeld was appointed professor at the University of Munich in 1906. He had studied mathematics and later shifted to mathematical physics. Sommerfeld became one of the most important pioneers of atomic and quantum physics, about which he published a first fundamental book in 1919. He was an excellent academic teacher and formed an influential school of theoretical physicists. Because of his mastery of mathematical instruments, Sommerfeld was able to apply Einstein's special theory of relativity to different problems of physics and thus contributed to the establishment of the theory in the years 1907–1910. In this period, Sommerfeld and Einstein also met personally and discussed problems of the early quantum theory. Their scientific exchange continued in later years through extensive correspondence.

### Synge, John (1897–1995)

The Irish theoretical physicist John Synge was appointed professor at Trinity College Dublin in 1925 and also taught at several universities in the United States and Canada. He made important contributions to different fields, including classical mechanics, geometric optics, and hydrodynamics, as well as mathematical physics and differential geometry. After the Second World War, Synge played a major role in the renewal and development of studies on the relativity theory, about which he published several books.

Photo: © Godfrey Argent Studio

### von Laue, Max (1879–1960)

The German theoretical physicist Max von Laue was appointed lecturer at the University of Berlin in 1906. From 1914 to 1919, he was professor at the University of Frankfurt, and from 1919, at the University of Berlin. Laue was particularly concerned with mathematical aspects of problems of optics and in 1907 delivered a mathematical explanation of a problem of light propagation in the frame of Einstein's special theory of relativity. Laue's work fostered the acceptance of the theory. He also contributed to its development by developing relativistic continuum mechanics. In 1912, he discovered the diffraction of X-rays by crystals. After their first meeting in 1906, Laue and Einstein became lifelong friends.

Photo: Bundesarchiv, Bild 183-U0205–502 / CC-BY-SA

### Weyl, Hermann (1885–1955)

In 1913, the German mathematician Hermann Weyl was appointed professor at the Technical University of Zurich, where he met Einstein and started to work on mathematical aspects of general relativity. In 1930, Weyl took over David Hilbert's position in Göttingen but immigrated to the United States in 1933, joining Einstein at the Institute for Advanced Study in Princeton. Weyl made important contributions to the development of theoretical physics and mathematics. His main concern in early years had been analysis and spectral theory, but he later became interested in topology and differential geometry. In 1918, Weyl published one of the most influential books on general relativity in which he offered new interpretations of some of its basic mathematical concepts. In the following years, he worked on the theory of groups and its application to quantum mechanics. In addition, Weyl concerned himself throughout his scientific career with the foundations of mathematics and philosophy of science.

# FURTHER READING

THE TOPICS COVERED IN THIS BOOK HAVE BEEN DEALT WITH EXTENSIVELY IN A RICH academic literature, ranging from physics to philosophy to the history of science. This collective effort of many scholars has produced a number of excellent comprehensive texts covering all aspects of the subject matter of this book. In the commentaries, we have made use of this vast intellectual resource. Here we have selected a short list of works based on two criteria: works that served us in our presentation of the history of general relativity, and those that may benefit nonspecialist readers who are interested in enriching their knowledge beyond the present exposition.

The authoritative edition of Einstein's papers is *The Collected Papers of Albert Einstein*, Vols. 1–14 (Princeton, NJ: Princeton University Press, 1987–). This edition contains numerous invaluable introductions to the various aspects of Einstein's biography and work. The translations given here are based on those featured in the English translation volumes. The published volumes of the Collected Papers have been made available online under einsteinpapers.press.princeton.edu. Also a substantial part of the Einstein Archives at The Hebrew University of Jerusalem are available online under www.alberteinstein.info.

Einstein himself endeavored to present his thoughts and theories in a generally understandable way. The corresponding publications are still among the most readable introductions to his work.

Einstein, Albert. 1920. *Relativity: The Special and the General Theory; A Popular Exposition by Albert Einstein*. Translated by R. W. Lawson. London: Methuen.

———. 1922. *The Meaning of Relativity: Four Lectures Delivered at Princeton University, May 1921 by Albert Einstein*. Translated by Edwin Plimpton Adams (1st ed.). London: Methuen.

———. 1992. *Autobiographical Notes. A Centennial Edition*. Edited by Paul A. Schilpp. La Salle, IL: Open Court.

———. 1950. *Out of My Later Years*. New York: Philosophical Library.

———. 1954. *Ideas and Opinions*. Based on *Mein Weltbild*, edited by Carl Seelig, and other sources. New translations and revisions by Sonja Bargmann. New York: Crown.

Einstein, Albert, and Leopold Infeld. 1938. *The Evolution of Physics: The Growth of Ideas from Early Concepts to Relativity and Quanta*. New York: Simon & Schuster.

A comprehensive collective work on the history of general relativity is:

Renn, Jürgen (ed.). 2007. *The Genesis of General Relativity*, 4 vols. Dordrecht: Springer.
Vol. 1: Jürgen Renn, Michel Janssen, John Norton, Tilman Sauer, and John Stachel. *Einstein's Zurich Notebook: Introduction and Source*.
Vol. 2: Jürgen Renn, Michel Janssen, John Norton, Tilman Sauer and John Stachel. *Einstein's Zurich Notebook: Commentary and Essays*.
Vol. 3: Jürgen Renn and Matthias Schemmel (eds.). *Gravitation in the Twilight of Classical Physics. Between Mechanics, Field Theory, and Astronomy*.
Vol. 4: Jürgen Renn and Matthias Schemmel (eds.). *Gravitation in the Twilight of Classical Physics. The Promise of Mathematics*.

An accessible, comprehensive, and up-to-date account of Einstein's scientific work is:

Lehner, Christoph, and Michel Janssen (eds.). 2014. *The Cambridge Companion to Einstein*. Cambridge: Cambridge University Press.

Important contributions on the history of general relativity have been published in the numerous volumes of the following series:

Howard, Don, and John Stachel (series eds.). 1989–. *Einstein Studies*. Boston: Birkhäuser/Springer.

A classic compilation of original studies on Einstein's life and science is:

Stachel, John. 2002. *Einstein from 'B' to 'Z'.* Vol. 9 of *Einstein Studies*. Boston: Birkhäuser.

There are many biographies on Einstein. Here, we have selected four that give prominence to the story told in this book:

Fölsing, Albrecht. 1997. *Albert Einstein: A Biography*. New York: Viking.
Isaacson, Walter. 2007. *Einstein: His Life and Universe*. New York: Simon & Schuster.
Neffe, Jürgen. 2007. *Einstein. A Biography*. New York: Farrar, Straus and Giroux.
Pais, Abraham. 1982. *'Subtle is the Lord . . .': The Science and the Life of Albert Einstein*. Oxford: Oxford University Press.

Important aspects of our story have become the subject of book-length studies that are particularly accessible to a general audience. The following are a few of them:

Eisenstaedt, Jean. 2006. *Curious History of Relativity: How Einstein's Theory of Gravity was Lost and Found Again*. Princeton, NJ: Princeton University Press.

Galison, Peter. 2003. *Einstein's Clocks, Poincare's Maps: Empires of Time*. New York: Norton.

Hentschel, Klaus. 1997. *The Einstein Tower: An Intertexture of Dynamic Construction, Relativity Theory, and Astronomy*. Palo Alto, CA: Stanford University Press.

Kennefick, Daniel. 2007. *Traveling at the Speed of Thought: Einstein and the Quest for Gravitational Waves*. Princeton, NJ: Princeton University Press.

Kragh, Helge. 1996. *Cosmology and Controversy: The Historical Development of Two Theories of the Universe*. Princeton, NJ: Princeton University Press.

Staley, Richard . 2008. *Einstein's Generation: The Origins of the Relativity Revolution*. Chicago: University of Chicago Press

Schutz, Bernard. 2004. *Gravity from the Ground Up*. Cambridge: Cambridge University Press.

Thorne, Kip S. 1994. *Black Holes and Time Warps: Einstein's Outrageous Legacy*. New York: Norton.

van Dongen, Jeroen. 2010. *Einstein's Unification*. Cambridge: Cambridge University Press.

Wazeck, Milena. 2014. *Einstein's Opponents: The Public Controversy about the Theory of Relativity in the 1920s*. Cambridge: Cambridge University Press.

The catalogues accompanying the Berlin exhibition "Albert Einstein Chief Engineer of the Universe" offer a comprehensive and opulently illustrated documentation of his life and work:

Renn, Jürgen. 2005. *Albert Einstein Chief Engineer of the Universe: Einstein's Life and Work in Context*. Berlin: Wiley-VCH.

———. *Albert Einstein Chief Engineer of the Universe: 100 Authors for Einstein*. Berlin: Wiley-VCH.

———. *Albert Einstein Chief Engineer of the Universe: Documents of a Life's Pathway*. Berlin: Wiley-VCH.

In addition, the following are several compilations of studies showing how broadly Einstein's work shaped modern science and culture:

Galison, Peter, Gerald James Holton, and Silvan S. Schweber (eds.). 2008. *Einstein for the 21st Century: His Legacy in Science, Art, and Modern Culture*. Princeton, NJ: Princeton University Press.

Schilpp, Paul Arthur (ed.). 1970. *Einstein, Albert: Philosopher-Scientist*. La Salle, IL: Open Court.

# THE FOUNDATION OF THE GENERAL THEORY OF RELATIVITY

By A. Einstein

The theory which is presented in the following pages conceivably constitutes the farthest-reaching generalization of a theory which, today, is generally called the "theory of relativity"; I will call the latter one—in order to distinguish it from the first named—the "special theory of relativity," which I assume to be known. The generalization of the theory of relativity has been facilitated considerably by Minkowski, a mathematician who was the first one to recognize the formal equivalence of space coordinates and the time coordinate, and utilized this in the construction of the theory. The mathematical tools that are necessary for general relativity were readily available in the "absolute differential calculus," which is based upon the research on non-Euclidean manifolds by Gauss, Riemann, and Christoffel, and which has been systematized by Ricci and Levi-Civita and has already been applied to problems of theoretical physics. In section B of the present paper I developed all the necessary mathematical tools—which cannot be assumed to be known to every physicist—and I tried to do it in as simple and transparent a manner as possible, so that a special study of the mathematical literature is not required for the understanding of the present paper. Finally, I want to acknowledge gratefully my friend, the mathematician Grossmann, whose help not only saved me the effort of studying the pertinent mathematical literature, but who also helped me in my search for the field equations of gravitation.

The following two translations are reproduced from the English edition of the *Collected Papers of Albert Einstein* (Doc. 30 and Doc. 41, vol. 6). The first paragraph of the first document and the entire second document have been translated by Alfred Engel; the remaining text of the first document is reprinted from H. A. Lorentz et al., *The Principle of Relativity*, trans. W. Perrett and G. B. Jeffery (Methuen, 1923; Dover rpt., 1952).

## A. FUNDAMENTAL CONSIDERATIONS ON THE POSTULATE OF RELATIVITY

### § I. Observations on the Special Theory of Relativity

THE SPECIAL THEORY OF RELATIVITY IS BASED ON THE FOLLOWING POSTULATE, WHICH IS also satisfied by the mechanics of Galileo and Newton.

If a system of co-ordinates K is chosen so that, in relation to it, physical laws hold good in their simplest form, the *same* laws also hold good in relation to any other system of co-ordinates K′ moving in uniform translation relatively to K. This postulate we call the "special principle of relativity." The word "special" is meant to intimate that the principle is restricted to the case when K′ has a motion of uniform translation relatively to K, but that the equivalence of K′ and K does not extend to the case of non-uniform motion of K′ relatively to K.

Thus the special theory of relativity does not depart from classical mechanics through the postulate of relativity, but through the postulate of the constancy of the velocity of light *in vacuo*, from which, in combination with the special principle of relativity, there follow, in the well-known way, the relativity of simultaneity, the Lorentzian transformation, and the related laws for the behavior of moving bodies and clocks.

The modification to which the special theory of relativity has subjected the theory of space and time is indeed far-reaching, but one important point has remained unaffected. For the laws of geometry, even according to the special theory of relativity, are to be interpreted directly as laws relating to the possible relative positions of solid bodies at rest; and, in a more general way, the laws of kinematics are to be interpreted as laws which describe the relations of measuring bodies and clocks. To two selected material points of a stationary rigid body there always corresponds a distance of quite definite length, which is independent of the locality and orientation of the body, and is also independent of the time. To two selected positions of the hands of a clock at rest relatively to the privileged system of reference there always corresponds an interval of time of a definite length, which is independent of place and time. We shall soon see that the general theory of relativity cannot adhere to this simple physical interpretation of space and time.

### § 2. The Need for an Extension of the Postulate of Relativity

In classical mechanics, and no less in the special theory of relativity, there is an inherent epistemological defect which was, perhaps for the first time, clearly pointed out by Ernst Mach. We will elucidate it by the following example:—Two fluid bodies of the same size and nature hover freely in space at so great a distance from each other and from all other masses that only those gravitational forces need be taken into account which arise from the interaction of different parts of the same body. Let the distance between the two bodies be invariable, and in neither of the bodies let there be any relative movements of the parts with respect to one another. But let either mass, as judged by an observer at rest relatively to the other mass, rotate with constant angular velocity about the line joining the masses. This is a verifiable relative motion of the two bodies. Now let us imagine that each of the bodies has been surveyed by means of measuring instruments at rest relatively to itself, and let the surface of $S_1$ prove to be a sphere, and that of $S_2$ an ellipsoid of revolution.

Thereupon we put the question—What is the reason for this difference in the two bodies? No answer can be admitted as epistemologically satisfactory,[1] unless the reason given is an *observable fact of experience*. The law of causality has not the significance of a statement as to the world of experience, except when *observable facts* ultimately appear as causes and effects.

Newtonian mechanics does not give a satisfactory answer to this question. It pronounces as follows:—The laws of mechanics apply to the space $R_1$, in respect to which the body $S_1$ is at rest, but not to the space $R_2$, in respect to which the body $S_2$ is at rest. But the privileged space $R_1$ of Galileo, thus introduced, is a merely *factitious* cause, and not a thing that can be observed. It is therefore clear that Newton's mechanics does not really satisfy the requirement of causality in the case under consideration, but only apparently does so, since it makes the factitious cause $R_1$ responsible for the observable difference in the bodies $S_1$ and $S_2$.

The only satisfactory answer must be that the physical system consisting of $S_1$ and $S_2$ reveals within itself no imaginable cause to which the differing behavior of $S_1$ and $S_2$ can be referred. The cause must therefore lie *outside* this system. We have to take it that the general laws of motion, which in particular determine the shapes of $S_1$ and $S_2$, must be such that the mechanical behavior of $S_1$ and $S_2$ is partly conditioned, in quite essential respects, by distant masses which we have not included in the system under consideration. These distant masses and their motions relative to $S_1$ and $S_2$ must then be regarded as the seat of the causes (which must be susceptible to observation) of the different behavior of our two bodies $S_1$ and $S_2$. They take over the rôle of the factitious cause $R_1$. Of all imaginable spaces $R_1$, $R_2$, etc., in any kind of motion relatively to one another, there is none which we may look upon as privileged *a priori* without reviving the above-mentioned epistemological objection. *The laws of physics must be of such a nature that they apply to systems of reference in any kind of motion.* Along this road we arrive at an extension of the postulate of relativity.

In addition to this weighty argument from the theory of knowledge, there is a well-known physical fact which favors an extension of the theory of relativity. Let K be a Galilean system of reference, i.e. a system relatively to which (at least in the four-dimensional region under consideration) a mass, sufficiently distant from other masses, is moving with uniform motion in a straight line. Let K′ be a second system of reference which is moving relatively to K in *uniformly accelerated* translation. Then, relatively to K′, a mass sufficiently distant from other masses would have an accelerated motion such that its acceleration and direction of acceleration are independent of the material composition and physical state of the mass.

Does this permit an observer at rest relatively to K′ to infer that he is on a "really" accelerated system of reference? The answer is in the negative; for the above-mentioned relation of freely movable masses to K′ may be interpreted equally well in the following way. The system of reference K′ is unaccelerated, but the space-time territory in question

---

1 Of course an answer may be satisfactory from the point of view of epistemology, and yet be unsound physically, if it is in conflict with other experiences.

is under the sway of a gravitational field, which generates the accelerated motion of the bodies relatively to K′.

This view is made possible for us by the teaching of experience as to the existence of a field of force, namely, the gravitational field, which possesses the remarkable property of imparting the same acceleration to all bodies.[2] The mechanical behavior of bodies relatively to K′ is the same as presents itself to experience in the case of systems which we are wont to regard as "stationary" or as "privileged." Therefore, from the physical standpoint, the assumption readily suggests itself that the systems K and K′ may both with equal right be looked upon as "stationary," that is to say, they have an equal title as systems of reference for the physical description of phenomena.

It will be seen from these reflections that in pursuing the general theory of relativity we shall be led to a theory of gravitation, since we are able to "produce" a gravitational field merely by changing the system of co-ordinates. It will also be obvious that the principle of the constancy of the velocity of light *in vacuo* must be modified, since we easily recognize that the path of a ray of light with respect to K′ must in general be curvilinear, if with respect to K light is propagated in a straight line with a definite constant velocity.

### § 3. The Space-Time Continuum. Requirement of General Co-Variance for the Equations Expressing General Laws of Nature

In classical mechanics, as well as in the special theory of relativity, the co-ordinates of space and time have a direct physical meaning. To say that a point-event has the $X_1$ coordinate $x_1$ means that the projection of the point-event on the axis of $X_1$, determined by rigid rods and in accordance with the. rules of Euclidean geometry, is obtained by measuring off a given rod (the unit of length) $x_1$ times from the origin of coordinates along the axis of $X_1$. To say that a point-event has the $X_4$ co-ordinate $x_4 = t$, means that a standard clock, made to measure time in a definite unit period, and which is stationary relatively to the system of co-ordinates and practically coincident in space with the point-event,[3] will have measured off $x_4 = t$ periods at the occurrence of the event.

This view of space and time has always been in the minds of physicists, even if, as a rule, they have been unconscious of it. This is clear from the part which these concepts play in physical measurements; it must also have underlain the reader's reflections on the preceding paragraph (§ 2) for him to connect any meaning with what he there read. But we shall now show that we must put it aside and replace it by a more general view, in order to be able to carry through the postulate of general relativity, if the special theory of relativity applies to the special case of the absence of a gravitational field.

In a space which is free of gravitational fields we introduce a Galilean system of reference K (*x, y, z, t*), and also a system of co-ordinates K′ (*x′, y′, z′, t′*) in uniform rotation relatively to K. Let the origins of both systems, as well as their axes of Z, permanently

---

2   Eötvös has proved experimentally that the gravitational field has this property in great accuracy.

3   We assume the possibility of verifying "simultaneity" for events immediately proximate in space, or—to speak more precisely—for immediate proximity or coincidence in space-time, without giving a definition of this fundamental concept.

coincide. We shall show that for a space-time measurement in the system K′ the above definition of the physical meaning of lengths and times cannot be maintained. For reasons of symmetry it is clear that a circle around the origin in the X, Y plane of K may at the same time be regarded as a circle in the X′, Y′ plane of K′. We suppose that the circumference and diameter of this circle have been measured with a unit measure infinitely small compared with the radius, and that we have the quotient of the two results. If this experiment were performed with a measuring-rod at rest relatively to the Galilean system K, the quotient would be π. With a measuring-rod at rest relatively to K′, the quotient would be greater than π. This is readily understood if we envisage the whole process of measuring from the "stationary" system K, and take into consideration that the measuring-rod applied to the periphery undergoes a Lorentzian contraction, while the one applied along the radius does not. Hence Euclidean geometry does not apply to K′. The notion of co-ordinates defined above, which presupposes the validity of Euclidean geometry, therefore breaks down in relation to the system K′. So, too, we are unable to introduce a time corresponding to physical requirements in K′, indicated by clocks at rest relatively to K′. To convince ourselves of this impossibility, let us imagine two clocks of identical constitution placed, one at the origin of co-ordinates, and the other at the circumference of the circle, and both envisaged from the "stationary" system K. By a familiar result of the special theory of relativity, the clock at the circumference—judged from K—goes more slowly than the other, because the former is in motion and the latter at rest. An observer at the common origin of co-ordinates, capable of observing the clock at the circumference by means of light, would therefore see it lagging behind the clock beside him. As he will not make up his mind to let the velocity of light along the path in question depend explicitly on the time, he will interpret his observations as showing that the clock at the circumference "really" goes more slowly than the clock at the origin. So he will be obliged to define time in such a way that the rate of a clock depends upon where the clock may be.

We therefore reach this result:—In the general theory of relativity, space and time cannot be defined in such a way that differences of the spatial co-ordinates can be directly measured by the unit measuring-rod, or differences in the time co-ordinate by a standard clock.

The method hitherto employed for laying co-ordinates into the space-time continuum in a definite manner thus breaks down, and there seems to be no other way which would allow us to adapt systems of co-ordinates to the four-dimensional universe so that we might expect from their application a particularly simple formulation of the laws of nature. So there is nothing for it but to regard all imaginable systems of co-ordinates, on principle, as equally suitable for the description of nature. This comes to requiring that:—

*The general laws of nature are to be expressed by equations which hold good for all systems of co-ordinates, that is, are co-variant with respect to any substitutions whatever (generally co-variant).*

It is clear that a physical theory which satisfies this postulate will also be suitable for the general postulate of relativity. For the sum of *all* substitutions in any case includes those which correspond to all relative motions of three-dimensional systems of co-ordinates. That this requirement of general co-variance, which takes away from space and time the last remnant of physical objectivity, is a natural one, will be seen from the following reflection. All our space-time verifications invariably amount to a determination of space-time coincidences. If, for example, events consisted merely in the motion of material points, then ultimately nothing would be observable but the meetings of two or more of these points. Moreover, the results of our measurings are nothing but verifications of such meetings of the material points of our measuring instruments with other material points, coincidences between the hands of a clock and points on the clock dial, and observed point-events happening at the same place at the same time.

The introduction of a system of reference serves no other purpose than to facilitate the description of the totality of such coincidences. We allot to the universe four space-time variables $x_1, x_2, x_3, x_4$ in such a way that for every point-event there is a corresponding system of values of the variables $x_1 \ldots x_4$. To two coincident point-events there corresponds one system of values of the variables $x_1 \ldots x_4$, i.e. coincidence is characterized by the identity of the co-ordinates. If, in place of the variables $x_1 \ldots x_4$, we introduce functions of them, $x'_1, x'_2, x'_3, x'_4$, as a new system of co-ordinates, so that the systems of values are made to correspond to one another without ambiguity, the equality of all four co-ordinates in the new system will also serve as an expression for the space-time coincidence of the two point-events. As all our physical experience can be ultimately reduced to such coincidences, there is no immediate reason for preferring certain systems of co-ordinates to others, that is to say, we arrive at the requirement of general co-variance.

### § 4. The Relation of the Four Co-ordinates to Measurement in Space and Time

It is not my purpose in this discussion to represent the general theory of relativity as a system that is as simple and logical as possible, and with the minimum number of axioms; but my main object is to develop this theory in such a way that the reader will feel that the path we have entered upon is psychologically the natural one, and that the underlying assumptions will seem to have the highest possible degree of security. With this aim in view let it now be granted that:—

For infinitely small four-dimensional regions the theory of relativity in the restricted sense is appropriate, if the coordinates are suitably chosen.

For this purpose we must choose the acceleration of the infinitely small ("local") system of co-ordinates so that no gravitational field occurs; this is possible for an infinitely small region. Let $X_1$, $X_2$, $X_3$, be the co-ordinates of space, and $X_4$ the appertaining co-ordinate of time measured in the appropriate unit.[4] If a rigid rod is imagined to be given as the unit measure, the co-ordinates, with a given orientation of the system of co-ordinates, have a direct physical meaning in the sense of the special theory of relativity. By the special theory of relativity the expression

$$ds^2 = -d\mathrm{X}_1^2 - d\mathrm{X}_2^2 - d\mathrm{X}_3^2 + d\mathrm{X}_4^2 \tag{1}$$

then has a value which is independent of the orientation of the local system of co-ordinates, and is ascertainable by measurements of space and time. The magnitude of the

4   The unit of time is to be chosen so that the velocity of light *in vacuo* as measured in the "local" system of co-ordinates is to be equal to unity.

linear element pertaining to points of the four-dimensional continuum in infinite prox-imity, we call $ds$. If the $ds$ belonging to the element $dX_1 \ldots dX_4$ is positive, we follow Min-kowski in calling it time-like; if it is negative, we call it space-like.

To the "linear element" in question, or to the two infinitely .proximate point-events, there will also correspond definite differentials $dx_1 \ldots dx_4$ of the four-dimensional co-or-dinates of any chosen system of reference. If this system, as well as the "local" system, is given for the region under consideration, the $dX_v$ will allow themselves to be represented here by definite linear homogeneous expressions of the $dx_\sigma$:—

$$dX_v = \sum_\sigma a_{v\sigma} dx_\sigma \qquad (2)$$

Inserting these expressions in (1), we obtain

$$ds^2 = \sum_{\tau\sigma} g_{\sigma\tau} dx_\sigma dx_\tau \qquad (3)$$

where the $g_{\sigma\tau}$ will be functions of the $x_\sigma$. These can no longer be dependent on the orien-tation and the state of motion of the "local" system of co-ordinates, for $ds^2$ is a quantity ascertainable by rod-clock measurement of point-events infinitely proximate in space-time, and defined independently of any particular choice of co-ordinates. The $g_{\sigma\tau}$ are to be chosen here so that $g_{\sigma\tau} = g_{\tau\sigma}$; the summation is to extend over all values of $\sigma$ and $\tau$, so that the sum consists of $4 \times 4$ terms, of which twelve are equal in pairs.

The case of the ordinary theory of relativity arises out of the case here considered, if it is possible, by reason of the particular relations of the $g_{\sigma\tau}$ in a finite region, to choose the sys-tem of reference in the finite region in such a way that the $g_{\sigma\tau}$ assume the constant values

$$\left. \begin{array}{cccc} -1 & 0 & 0 & 0 \\ 0 & -1 & 0 & 0 \\ 0 & 0 & -1 & 0 \\ 0 & 0 & 0 & +1 \end{array} \right\} \qquad (4)$$

We shall find hereafter that the choice of such co-ordinates is, in general, not possible for a finite region.

From the considerations of § 2 and § 3 it follows that the quantities $g_{\tau\sigma}$ are to be regarded from the physical standpoint as the quantities which describe the gravitational field in relation to the chosen system of reference. For, if we now assume the special the-ory of relativity to apply to a certain four-dimensional region with the co-ordinates prop-erly chosen, then the $g_{\sigma\tau}$ have the values given in (4). A free material point then moves, relatively to this system, with uniform motion in a straight line. Then if we introduce new space-time co-ordinates $x_1, x_2, x_3, x_4$, by means of any substitution we choose, the $g_{\sigma\tau}$ in this new system will no longer be constants, but functions of space and time. At the same time the motion of the free material point will present itself in the new co-ordinates as a curvilinear non-uniform motion, and the law of this motion will be independent of the nature of the moving particle. We shall therefore interpret this motion as a motion under the influence of a gravitational field. We thus find the occurrence of a gravitational field connected with a space-time variability of the $g_\sigma$. So, too, in the general case, when we are no longer able by a suitable choice of co-ordinates to apply the special theory of relativity to a finite region, we shall hold fast to the view that the $g_{\sigma\tau}$ describe the gravitational field.

Thus, according to the general theory of relativity, gravitation occupies an exceptional position with regard to other forces, particularly the electromagnetic forces, since the ten functions representing the gravitational field at the same time define the metrical properties of the space measured.

## B. MATHEMATICAL AIDS TO THE FORMULATION OF GENERALLY COVARIANT EQUATIONS

Having seen in the foregoing that the general postulate of relativity leads to the requirement that the equations of physics shall be covariant in the face of any substitution of the co-ordinates $x_1 \ldots x_4$, we have to consider how such generally covariant equations can be found. We now turn to this purely mathematical task, and we shall find that in its solution a fundamental rôle is played by the invariant $ds$ given in equation (3), which, borrowing from Gauss's theory of surfaces, we have called the "linear element."

The fundamental idea of this general theory of covariants is the following:—Let certain things ("tensors") be defined with respect to any system of co-ordinates by a number of functions of the co-ordinates, called the "components" of the tensor. There are then certain rules by which these components can be calculated for a new system of co-ordinates, if they are known for the original system of co-ordinates, and if the transformation connecting the two systems is known. The things hereafter called tensors are further characterized by the fact that the equations of transformation for their components are linear and homogeneous. Accordingly, all the components in the new system vanish, if they all vanish in the original system. If, therefore, a law of nature is expressed by equating all the components of a tensor to zero, it is generally covariant. By examining the laws of the formation of tensors, we acquire the means of formulating generally covariant laws.

### § 5. Contravariant and Covariant Four-vectors

*Contravariant Four-vectors.*—The linear element is defined by the four "components" $dx_\nu$, for which the law of transformation is expressed by the equation

$$dx'_\sigma = \sum_\nu \frac{\partial x'_\sigma}{\partial x_\nu} dx_\nu \qquad (5)$$

The $dx'_\sigma$ are expressed as linear and homogeneous functions of the $dx_\nu$. Hence we may look upon these co-ordinate differentials as the components of a "tensor" of the particular kind which we call a contravariant four-vector. Any thing which is defined relatively to the system of co-ordinates by four quantities $A^\nu$, and which is transformed by the same law

$$A'^\sigma = \sum_\nu \frac{\partial x'_\sigma}{\partial x_\nu} A^\nu, \qquad (5a)$$

we also call a contravariant four-vector. From (5a) it follows at once that the sums $A^\sigma \pm B^\sigma$ are also components of a four-vector, if $A^\sigma$ and $B^\sigma$ are such. Corresponding relations hold for all "tensors" subsequently to be introduced. (Rule for the addition and subtraction of tensors.)

*Covariant Four-vectors.*—We call four quantities $A_\nu$ the components of a covariant four-vector, if for any arbitrary choice of the contravariant four-vector $B^\nu$

$$\sum_\nu A_\nu B^\nu = \text{Invariant} \tag{6}$$

The law of transformation of a covariant four-vector follows from this definition. For if we replace $B^\nu$ on the right-hand side of the equation

$$\sum_\sigma A'_\sigma B'^\sigma = \sum_\nu A_\nu B^\nu$$

by the expression resulting from the inversion of (5a),

$$\sum_\sigma \frac{\partial x_\nu}{\partial x'_\sigma} B'^\sigma,$$

we obtain

$$\sum_\sigma B'^\sigma \sum_\nu \frac{\partial x_\nu}{\partial x'_\sigma} A_\nu = \sum_\sigma B'^\sigma A'_\sigma.$$

Since this equation is true for arbitrary values of the $B'^\sigma$, it follows that the law of transformation is

$$A'_\sigma = \sum_\nu \frac{\partial x_\nu}{\partial x'_\sigma} A_\nu \tag{7}$$

*Note on a Simplified Way of Writing the Expressions.*—A glance at the equations of this paragraph shows that there is always a summation with respect to the indices which occur twice under a sign of summation (e.g. the index $\nu$ in (5)), and only with respect to indices which occur twice. It is therefore possible, without loss of clearness, to omit the sign of summation. In its place we introduce the convention:— If an index occurs twice in one term of an expression, it is always to be summed unless the contrary is expressly stated.

The difference between covariant and contravariant four-vectors lies in the law of transformation ((7) or (5) respectively). Both forms are tensors in the sense of the general remark above. Therein lies their importance. Following Ricci and Levi-Civita, we denote the contravariant character by placing the index above, the covariant by placing it below.

## § 6. Tensors of the Second and Higher Ranks

*Contravariant Tensors.*—If we form all the sixteen products $A^{\mu\nu}$ of the components $A^\mu$ and $B^\nu$ of two contravariant four-vectors

$$A^{\mu\nu} = A^\mu B^\nu \tag{8}$$

then by (8) and (5a) $A^{\mu\nu}$ satisfies the law of transformation

$$A'^{\sigma\tau} = \frac{\partial x'_\sigma}{\partial x_\mu}\frac{\partial x'_\tau}{\partial x_\nu}A^{\mu\nu} \tag{9}$$

We call a thing which is described relatively to any system of reference by sixteen quantities, satisfying the law of transformation (9), a contravariant tensor of the second rank. Not every such tensor allows itself to be formed in accordance with (8) from two four-vectors, but it is easily shown that any given sixteen $A^{\mu\nu}$ can be represented as the sums of the $A^{\mu\nu}$ of four appropriately selected pairs of four-vectors. Hence we can prove nearly all the laws which apply to the tensor of the second rank defined by (9) in the simplest manner by demonstrating them for the special tensors of the type (8).

*Contravariant Tensors of Any Rank.*—It is clear that, on the lines of (8) and (9), contravariant tensors of the third and higher ranks may also be defined with $4^3$ components, and so on. In the same way it follows from (8) and (9) that the contravariant four-vector may be taken in this sense as a contravariant tensor of the first rank.

*Covariant Tensors.*—On the other hand, if we take the sixteen products $A_{\mu\nu}$ of two covariant four-vectors $A_\mu$ and $B_\nu$

$$A_{\mu\nu} = A_\mu B_\nu, \tag{10}$$

the law of transformation for these is

$$A'_{\sigma\tau} = \frac{\partial x_\mu}{\partial x'_\sigma}\frac{\partial x_\nu}{\partial x'_\tau}A_{\mu\nu} \tag{11}$$

This law of transformation defines the covariant tensor of the second rank. All our previous remarks on contravariant tensors apply equally to covariant tensors.

NOTE.—It is convenient to treat the scalar (or invariant) both as a contravariant and a covariant tensor of zero rank.

*Mixed Tensors.*—We may also define a tensor of the second rank of the type

$$A_\mu^\nu = A_\mu B^\nu \tag{12}$$

which is covariant with respect to the index $\mu$, and contravariant with respect to the index $\nu$. Its law of transformation is

$$A'^\tau_\sigma = \frac{\partial x'_\tau}{\partial x_\nu}\frac{\partial x_\mu}{\partial x'_\sigma}A_\mu^\nu \tag{13}$$

Naturally there are mixed tensors with any number of indices of covariant character, and any number of indices of contravariant character. Covariant and contravariant tensors may be looked upon as special cases of mixed tensors.

*Symmetrical Tensors.*—A contravariant, or a covariant tensor, of the second or higher rank is said to be symmetrical if two components, which are obtained the one from the other by the interchange of two indices, are equal. The tensor $A^{\mu\nu}$, or the tensor $A_{\mu\nu}$, is thus symmetrical if for any combination of the indices $\mu$, $\nu$

$$A^{\mu\nu} = A^{\nu\mu}, \qquad (14)$$

or respectively,

$$A_{\mu\nu} = A_{\nu\mu} \qquad (14a)$$

It has to be proved that the symmetry thus defined is a property which is independent of the system of reference. It follows in fact from (9), when (14) is taken into consideration, that

$$A'^{\sigma\tau} = \frac{\partial x'_\sigma}{\partial x_\mu} \frac{\partial x'_\tau}{\partial x_\nu} A^{\mu\nu} = \frac{\partial x'_\sigma}{\partial x_\mu} \frac{\partial x'_\tau}{\partial x_\nu} A^{\nu\mu} = \frac{\partial x'_\sigma}{\partial x_\nu} \frac{\partial x'_\tau}{\partial x_\mu} A^{\mu\nu} = A'^{\tau\sigma}.$$

The last equation but one depends upon the interchange of the summation indices $\mu$ and $\nu$, i.e. merely on a change of notation.

*Antisymmetrical Tensors.*—A contravariant or a covariant tensor of the second, third, or fourth rank is said to be anti-symmetrical if two components, which are obtained the one from the other by the interchange of two indices, are equal and of opposite sign The tensor $A^{\mu\nu}$, or the tensor $A_{\mu\nu}$, is therefore antisymmetrical, if always

$$A^{\mu\nu} = -A^{\nu\mu}, \qquad (15)$$

or respectively,

$$A_{\mu\nu} = -A_{\nu\mu} \qquad (15a)$$

Of the sixteen components $A^{\mu\nu}$, the four components $A^{\mu\mu}$ vanish; the rest are equal and of opposite sign in pairs, so that there are only six components numerically different (a six-vector). Similarly we see that the antisymmetrical tensor of the third rank $A^{\mu\nu\sigma}$ has only four numerically different components, while the antisymmetrical tensor $A^{\mu\nu\sigma\tau}$ has only one. There are no antisymmetrical tensors of higher rank than the fourth in a continuum of four dimensions.

## § 7. Multiplication of Tensors

*Outer Multiplication of Tensors.*—We obtain from the components of a tensor of rank $n$ and of a tensor of rank $m$ the components of a tensor of rank $n + m$ by multiplying each component of the one tensor by each component of the other. Thus, for example, the tensors T arise out of the tensors A and B of different kinds,

$$T_{\mu\nu\sigma} = A_{\mu\nu}B_\sigma,$$
$$T^{\mu\nu\sigma\tau} = A^{\mu\nu}B^{\sigma\tau},$$
$$T^{\sigma\tau}_{\mu\nu} = A_{\mu\nu}B_{\sigma\nu}.$$

The proof of the tensor character of T is given directly by the representations (8), (10), (12), or by the laws of transformation (9), (11), (13). The equations (8), (10), (12) are themselves examples of outer multiplication of tensors of the first rank.

*"Contraction" of a Mixed Tensor.*—From any mixed tensor we may form a tensor whose rank is less by two, by equating an index of covariant with one of contravariant character, and summing with respect to this index ("contraction"). Thus, for example, from the mixed tensor of the fourth rank $A_{\mu\nu}^{\sigma\tau}$, we obtain the mixed tensor of the second rank,

$$A_{\nu}^{\tau} = A_{\mu\nu}^{\mu\tau}\left(= \sum_{\mu} A_{\mu\nu}^{\mu\tau}\right),$$

and from this, by a second contraction, the tensor of zero rank,

$$A = A_{\nu}^{\nu} = A_{\mu\nu}^{\mu\tau},$$

The proof that the result of contraction really possesses the tensor character is given either by the representation of a tensor according to the generalization of (12) in combination with (6), or by the generalization of (13).

*Inner and Mixed Multiplication of Tensors.*—These consist in a combination of outer multiplication with contraction.

*Examples.*—From the covariant tensor of the second rank $A^{\mu\nu}$ and the contravariant tensor of the first rank $B^{\sigma}$ we form by outer multiplication the mixed tensor

$$D_{\mu\nu}^{\sigma} = A_{\mu\nu}B^{\sigma}.$$

On contraction with respect to the indices $\nu$ and $\sigma$, we obtain the covariant four-vector

$$D_{\mu} = D_{\mu\nu}^{\nu} = A_{\mu\nu}B^{\nu}.$$

This we call the inner product of the tensors $A_{\mu\nu}$ and $B^{\sigma}$. Analogously we form from the tensors $A_{\mu\nu}$ and $B^{\sigma\tau}$, by outer multiplication and double contraction, the inner product $A_{\mu\nu}B^{\mu\nu}$. By outer multiplication and one contraction, we obtain from $A_{\mu\nu}$ and $B^{\sigma\tau}$ the mixed tensor of the second rank $D_{\mu}^{\tau} = A_{\mu\nu}B^{\nu\tau}$. This operation may be aptly characterized as a mixed one, being "outer" with respect to the indices $\mu$ and $\tau$, and "inner" with respect to the indices $\nu$ and $\sigma$.

We now prove a proposition which is often useful as evidence of tensor character. From what has just been explained, $A_{\mu\nu}B^{\mu\nu}$ is a scalar if $A_{\mu\nu}$ and $B^{\sigma\tau}$ are tensors. But we may also make the following assertion: If $A_{\mu\nu}B^{\mu\nu}$ is a scalar *for any choice of the tensor* $B^{\mu\nu}$, then $A_{\mu\nu}$ has tensor character. For, by hypothesis, for any substitution,

$$A_{\sigma\tau}'B'^{\sigma\tau} = A_{\mu\nu}B^{\mu\nu}.$$

But by an inversion of (9)

$$B^{\mu\nu} = \frac{\partial x_{\mu}}{\partial x_{\sigma}'}\frac{\partial x_{\nu}}{\partial x_{\tau}'}B'^{\sigma\tau}.$$

This, inserted in the above equation, gives

$$\left(A'_{\sigma\tau} - \frac{\partial x_\mu}{\partial x'_\sigma}\frac{\partial x_\nu}{\partial x'_\tau}A_{\mu\nu}\right)B'^{\sigma\tau} = 0.$$

This can only be satisfied for arbitrary values of $B'^{\sigma\tau}$ if the bracket vanishes. The result then follows by equation (11). This rule applies correspondingly to tensors of any rank and character, and the proof is analogous in all cases.

The rule may also be demonstrated in this form: If $B^\mu$ and $C^\nu$ are any vectors, and if, for all values of these, the inner product $A_{\mu\nu}B^\mu C^\nu$ is a scalar, then $A_{\mu\nu}$ is a covariant tensor. This latter proposition also holds good even if only the more special assertion is correct, that with any choice of the four-vector $B^\mu$ the inner product $A_{\mu\nu}B^\mu B^\nu$ is a scalar, if in addition it is known that $A_{\mu\nu}$ satisfies the condition of symmetry $A_{\mu\nu} = A_{\nu\mu}$. For by the method given above we prove the tensor character of $(A_{\mu\nu} + A_{\nu\mu})$, and from this the tensor character of $A_{\mu\nu}$ follows on account of symmetry. This also can be easily generalized to the case of covariant and contravariant tensors of any rank.

Finally, there follows from what has been proved, this law, which may also be generalized for any tensors: If for any choice of the four-vector $B^\nu$ the quantities $A_{\mu\nu}B^\nu$ form a tensor of the first rank, then $A_{\mu\nu}$ is a tensor of the second rank. For, if $C^\mu$ is any four-vector, then on account of the tensor character of $A_{\mu\nu}B^\nu$, the inner product $A_{\mu\nu}B^\nu C^\mu$ is a scalar for any choice of the two four-vectors $B^\nu$ and $C^\mu$. From which the proposition follows.

## § 8. Some Aspects of the Fundamental Tensor

*The Covariant Fundamental Tensor.*—In the invariant expression for the square of the linear element,

$$ds^2 = g_{\mu\nu}dx_\mu dx_\nu,$$

the part played by the $dx_\mu$ is that of a contravariant vector which may be chosen at will. Since further, $g_{\mu\nu} = g_{\nu\mu}$, it follows from the considerations of the preceding paragraph that $g_{\mu\nu}$ is a covariant tensor of the second rank. We call it the "fundamental tensor." In what follows we deduce some properties of this tensor which, it is true, apply to any tensor of the second rank. But as the fundamental tensor plays a special part in our theory, which has its physical basis in the peculiar effects of gravitation, it so happens that the relations to be developed are of importance to us only in the case of the fundamental tensor.

*The Contravariant Fundamental Tensor.*—If in the determinant formed by the elements $g_{\mu\nu}$, we take the co-factor of each of the $g_{\mu\nu}$ and divide it by the determinant $g = |g_{\mu\nu}|$, we obtain certain quantities $g^{\mu\nu}(= g^{\nu\mu})$ which, as we shall demonstrate, form a contravariant tensor.

By a known property of determinants

$$g_{\mu\sigma}g^{\nu\sigma} = \delta_\mu^\nu \tag{16}$$

where the symbol $\delta_\mu^\nu$ denotes 1 or 0, according as $\mu = \nu$ or $\mu \neq \nu$.

Instead of the above expression for $ds^2$ we may thus write

$$g_{\mu\sigma} \delta_\nu^\sigma dx_\mu dx_\nu$$

or, by (16)

$$g_{\mu\sigma} g_{\nu\tau} g^{\sigma\tau} dx_\mu dx_\nu.$$

But, by the multiplication rules of the preceding paragraphs, the quantities

$$d\xi_\sigma = g_{\mu\sigma} dx_\mu$$

form a covariant four-vector, and in fact an arbitrary vector, since the $dx_\mu$ are arbitrary. By introducing this into our expression we obtain

$$ds^2 = g^{\sigma\tau} d\xi_\sigma d\xi_\tau.$$

Since this, with the arbitrary choice of the vector $d\xi_\sigma$, is a scalar, and $g^{\sigma\tau}$ by its definition is symmetrical in the indices $\sigma$ and $\tau$, it follows from the results of the preceding paragraph that $g^{\sigma\tau}$ is a contravariant tensor.

It further follows from (16) that $\delta_\mu$ is also a tensor, which we may call the mixed fundamental tensor.

*The Determinant of the Fundamental Tensor.*—By the rule for the multiplication of determinants

$$\left| g_{\mu\alpha} g^{\alpha\nu} \right| = \left| g_{\mu\alpha} \right| \times \left| g^{\alpha\nu} \right|.$$

On the other hand

$$\left| g_{\mu\alpha} g^{\alpha\nu} \right| = \left| \delta_\mu^\nu \right| = 1.$$

It therefore follows that

$$\left| g_{\mu\nu} \right| \times \left| g^{\mu\nu} \right| = 1 \tag{17}$$

*The Volume Scalar.*—We seek first the law of transformation of the determinant $g = \left| g_{\mu\nu} \right|$. In accordance with (11)

$$g' = \left| \frac{\partial x_\mu}{\partial x'_\sigma} \frac{\partial x}{\partial x'_\tau} g_{\mu\nu} \right|.$$

Hence, by a double application of the rule for the multiplication of determinants, it follows that

$$g' = \left| \frac{\partial x_\mu}{\partial x'_\sigma} \right| \cdot \left| \frac{\partial x_\nu}{\partial x'_\tau} \right| \cdot \left| g_{\mu\nu} \right| = \left| \frac{\partial x_\mu}{\partial x'_\sigma} \right|^2 g,$$

or

$$\sqrt{g'} = \left| \frac{\partial x_\mu}{\partial x'_\sigma} \right| \sqrt{g}.$$

On the other hand, the law of transformation of the element of volume

$$d\tau = \int dx_1 dx_2 dx_3 dx_4$$

is, in accordance with the theorem of Jacobi,

$$d\tau' = \left| \frac{\partial x'_\sigma}{\partial x_\mu} \right| d\tau.$$

By multiplication of the last two equations, we obtain

$$\sqrt{g'}\, d\tau' = \sqrt{g}\, d\tau \tag{18}.$$

Instead of $\sqrt{g}$, we introduce in what follows the quantity $\sqrt{-g}$, which is always real on account of the hyperbolic character of the space-time continuum. The invariant $\sqrt{-g}\, d\tau$ is equal to the magnitude of the four-dimensional element of volume in the "local" system of reference, as measured with rigid rods and clocks in the sense of the special theory of relativity.

*Note on the Character of the Space-time Continuum.*—Our assumption that the special theory of relativity can always be applied to an infinitely small region, implies that $ds^2$ can always be expressed in accordance with (1) by means of real quantities $dX_1 \ldots dX_4$. If we denote by $d\tau_0$ the "natural" element of volume $dX_1, dX_2, dX_3, dX_4$, then

$$d\tau_0 = \sqrt{-g}\, d\tau \tag{18a}$$

If $\sqrt{-g}$ were to vanish at a point of the four-dimensional continuum, it would mean that at this point an infinitely small "natural" volume would correspond to a finite volume in the co-ordinates. Let us assume that this is never the case. Then $g$ cannot change sign. We will assume that, in the sense of the special theory of relativity, $g$ always has a finite negative value. This is a hypothesis as to the physical nature of the continuum under consideration, and at the same time a convention as to the choice of co-ordinates.

But if $-g$ is always finite and positive, it is natural to settle the choice of co-ordinates *a posteriori* in such a way that this quantity is always equal to unity. We shall see later that by such a restriction of the choice of co-ordinates it is possible to achieve an important simplification of the laws of nature.

In place of (18), we then have simply $d\tau' = d\tau$, from which, in view of Jacobi's theorem, it follows that

$$\left| \frac{\partial x'_\sigma}{\partial x_\mu} \right| = 1 \tag{19}$$

Thus, with this choice of co-ordinates, only substitutions for which the determinant is unity are permissible.

But it would be erroneous to believe that this step indicates a partial abandonment of the general postulate of relativity. We do not ask "What are the laws of nature which are co-variant in face of all substitutions for which the determinant is unity?" but our question is "What are the generally co-variant laws of nature?" It is not until we have formulated these that we simplify their expression by a particular choice of the system of reference.

*The Formation of New Tensors by Means of the Fundamental Tensor.*—Inner, outer, and mixed multiplication of a tensor by the fundamental tensor give tensors of different character and rank. For example,

$$A^{\mu} = g^{u\sigma}A_{\sigma},$$
$$A = g_{\mu\nu}A^{\mu\nu}.$$

The following forms may be specially noted:—

$$A^{\mu\nu} = g^{\mu\alpha}g^{\nu\beta}A_{\alpha\beta},$$
$$A^{\mu\nu} = g_{\mu\alpha}g_{\nu\beta}A^{\alpha\beta}$$

(the "complements" of covariant and contravariant tensors respectively), and

$$B_{\mu\nu} = g_{\mu\nu}g^{\alpha\beta}A_{\alpha\beta}.$$

We call $B_{\mu\nu}$ the reduced tensor associated with $A_{\mu\nu}$. Similarly,

$$B^{\mu\nu} = g^{\mu\nu}g_{\alpha\beta}A^{\alpha\beta}.$$

It may be noted that $g^{\mu\nu}$ is nothing more than the complement of $g_{\mu\nu}$, since

$$g^{\mu\alpha}g^{\nu\beta}g_{\alpha\beta} = g^{\mu\alpha}\delta_{\alpha}^{\nu} = g^{\mu\nu}.$$

### § 9. The Equation of the Geodetic Line. The Motion of a Particle

As the linear element $ds$ is defined independently of the system of co-ordinates, the line drawn between two points P and P′ of the four-dimensional continuum in such a way that $\int ds$ is stationary—a geodetic line—has a meaning which also is independent of the choice of co-ordinates. Its equation is

$$\delta \int_{P}^{P'} ds = 0 \qquad\qquad (20)$$

Carrying out the variation in the usual way, we obtain from this equation four differential equations which define the geodetic line; this operation will be inserted here for the sake of completeness. Let $\lambda$ be a function of the co-ordinates $x_{\nu}$, and let this define a family of surfaces which intersect the required geodetic line as well as all the lines in immediate proximity to it which are drawn through the points P and P′. Any such line may then be supposed to be given by expressing its co-ordinates $x_v$ as functions of $\lambda$. Let the symbol $\delta$ indicate the transition from a point of the required geodetic to the point corresponding to the same $\lambda$ on a neighboring line. Then for (20) we may substitute

$$\left.\begin{aligned}
\int_{\lambda_1}^{\lambda_2} \delta w d\lambda &= 0 \\
w^2 &= g_{\mu\nu}\frac{dx_\mu}{d\lambda}\frac{dx_\nu}{d\lambda}
\end{aligned}\right\} \tag{20a}$$

But since

$$\delta w = \frac{1}{w}\left\{\frac{1}{2}\frac{\partial g_{\mu\nu}}{\partial x_\sigma}\frac{dx_\mu}{d\lambda}\frac{dx_\nu}{d\lambda}\delta x_\sigma + g_{\mu\nu}\frac{dx_\mu}{d\lambda}\delta\left(\frac{dx_\nu}{d\lambda}\right)\right\},$$

and

$$\delta\left(\frac{dx_\nu}{d\lambda}\right) = \frac{d}{d\lambda}(\delta x_\nu),$$

we obtain from (20a), after a partial integration,

$$\int_{\lambda_1}^{\lambda_2} k_\sigma \delta x_\sigma d\lambda = 0,$$

where

$$k_\sigma = \frac{d}{d\lambda}\left\{\frac{g_{\mu\nu}}{w}\frac{dx_\mu}{d\lambda}\right\} - \frac{1}{2w}\frac{\partial g_{\mu\nu}}{\partial x_\sigma}\frac{dx_\mu}{d\lambda}\frac{dx_\nu}{d\lambda} \tag{20b}$$

Since the values of $\delta x_\sigma$ are arbitrary, it follows from this that

$$k_\sigma = 0 \tag{20c}$$

are the equations of the geodetic line.

If $ds$ does not vanish along the geodetic line we may choose the "length of the arc" $s$, measured along the geodetic line, for the parameter $\lambda$. Then $w = 1$, and in place of (20c) we obtain

$$g_{\mu\nu}\frac{d^2x_\mu}{ds^2} + \frac{\partial g_{\mu\nu}}{\partial x_\sigma}\frac{dx_\sigma}{ds}\frac{dx_\mu}{ds} - \frac{1}{2}\frac{\partial g_{\mu\nu}}{\partial x_\sigma}\frac{dx_\mu}{ds}\frac{dx_\nu}{ds} = 0$$

or, by a mere change of notation,

$$g_{\alpha\sigma}\frac{d^2x_\alpha}{ds^2} + [\mu\nu,\sigma]\frac{dx_\mu}{ds}\frac{dx_\nu}{ds} = 0 \tag{20d}$$

where, following Christoffel, we have written

$$[\mu\nu,\sigma] = \frac{1}{2}\left(\frac{\partial g_{\mu\sigma}}{\partial x_\nu} + \frac{\partial g_{\nu\sigma}}{\partial x_\mu} - \frac{\partial g_{\mu\nu}}{\partial x_\sigma}\right) \tag{21}$$

Finally, if we multiply (20d) by $g^{\sigma\tau}$ (outer multiplication with respect to $\tau$, inner with respect to $\sigma$), we obtain the equations of the geodetic line in the form

$$\frac{d^2 x_\tau}{ds^2} + \{\mu\nu,\tau\}\frac{dx_\mu}{ds}\frac{dx_\nu}{ds} = 0 \tag{22}$$

where, following Christoffel, we have set

$$\{\mu\nu,\tau\} = g^{\tau\alpha}[\mu\nu,\alpha] \tag{23}$$

## § 10. The Formation of Tensors by Differentiation

With the help of the equation of the geodetic line we can now easily deduce the laws by which new tensors can be formed from old by differentiation. By this means we are able for the first time to formulate generally covariant differential equations. We reach this goal by repeated application of the following simple law:—

   If in our continuum a curve is given, the points of which are specified by the arcual distance $s$ measured from a fixed point on the curve, and if, further, $\phi$ is an invariant function of space, then $d\phi/ds$ is also an invariant. The proof lies in this, that $ds$ is an invariant as well as $d\phi$.

   As

$$\frac{d\phi}{ds} = \frac{\partial\phi}{\partial x_\mu}\frac{dx_\mu}{ds}$$

therefore

$$\psi = \frac{\partial\phi}{dx_\mu}\frac{dx_\mu}{ds}$$

is also an invariant, and an invariant for all curves starting from a point of the continuum, that is, for any choice of the vector $dx_\mu$. Hence it immediately follows that

$$A_\mu = \frac{\partial\phi}{\partial x_\mu} \tag{24}$$

is a covariant four-vector—the "gradient" of $\phi$.

   According to our rule, the differential quotient

$$\chi = \frac{d\psi}{ds}$$

taken on a curve, is similarly an invariant. Inserting the value of $\psi$, we obtain in the first place

$$\chi = \frac{\partial^2 \phi}{\partial x_\mu \partial x_\nu} \frac{dx_\mu}{ds} \frac{dx_\nu}{ds} + \frac{\partial \phi}{\partial x_\mu} \frac{d^2 x_\mu}{ds^2}.$$

The existence of a tensor cannot be deduced from this forthwith. But if we may take the curve along which we have differentiated to be a geodetic, we obtain on substitution for $d^2 x_\nu / ds^2$ from (22),

$$\chi = \left( \frac{\partial^2 \phi}{\partial x_\mu \partial x_\nu} - \{\mu\nu, \tau\} \frac{\partial \phi}{\partial x_\tau} \right) \frac{dx_\mu}{ds} \frac{dx_\nu}{ds}.$$

Since we may interchange the order of the differentiations, and since by (23) and (21) $\{\mu\nu, \tau\}$ is symmetrical in $\mu$ and $\nu$, it follows that the expression in brackets is symmetrical in $\mu$ and $\nu$. Since a geodetic line can be drawn in any direction from a point of the continuum, and therefore $dx_\mu / ds$ is a four-vector with the ratio of its components arbitrary, it follows from the results of § 7 that

$$A_{\mu\nu} = \frac{\partial^2 \phi}{\partial x_\mu \partial x_\nu} - \{\mu\nu, \tau\} \frac{\partial \phi}{\partial x_\tau} \tag{25}$$

is a covariant tensor of the second rank. We have therefore come to this result: from the covariant tensor of the first rank

$$A_\mu = \frac{\partial \phi}{\partial x_\mu}$$

we can, by differentiation, form a covariant tensor of the second rank

$$A_{\mu\nu} = \frac{\partial A_\mu}{\partial x_\nu} - \{\mu\nu, \tau\} A_\tau \tag{26}$$

We call the tensor $A_{\mu\nu}$ the "extension" (covariant derivative) of the tensor $A_\mu$. In the first place we can readily show that the operation leads to a tensor, even if the vector $A_\mu$ cannot be represented as a gradient. To see this, we first observe that

$$\psi \frac{\partial \phi}{\partial x_\mu}$$

is a covariant vector, if $\psi$ and $\phi$ are scalars. The sum of four such terms

$$S_\mu = \psi^{(1)} \frac{\phi \partial^{(1)}}{\partial x_\mu} + . + . + \psi^{(4)} \frac{\partial \phi^{(4)}}{\partial x_\mu},$$

is also a covariant vector, if $\psi^{(1)}$, $\phi^{(1)}$ ... $\psi^{(4)}$, $\phi^{(4)}$ are scalars. But it is clear that any covariant vector can be represented in the form $S_\mu$. For, if $A_\mu$ is a vector whose components are any given functions of the $x_v$, we have only to put (in terms of the selected system of co-ordinates)

$$\psi^{(1)} = A_1, \quad \phi^{(1)} = x_1,$$
$$\psi^{(2)} = A_2, \quad \phi^{(2)} = x_2,$$
$$\psi^{(3)} = A_3, \quad \phi^{(3)} = x_3,$$
$$\psi^{(4)} = A_4, \quad \phi^{(4)} = x_4,$$

in order to ensure that $S_\mu$ shall be equal to $A_\mu$.

Therefore, in order to demonstrate that $A_{\mu v}$ is a tensor if *any* covariant vector is inserted on the right-hand side for $A_\mu$, we only need show that this is so for the vector $S_\mu$. But for this latter purpose it is sufficient, as a glance at the right-hand side of (26) teaches us, to furnish the proof for the case

$$A_\mu = \psi \frac{\partial \phi}{\partial x_\mu}.$$

Now the right-hand side of (25) multiplied by $\psi$,

$$\psi \frac{\partial^2 \phi}{\partial x_\mu \partial x_v} - \{\mu v, \tau\} \psi \frac{\partial \phi}{\partial x_\tau}$$

is a tensor. Similarly

$$\frac{\partial \psi}{\partial x_\mu} \frac{\partial \phi}{\partial x_v}$$

being the outer product of two vectors, is a tensor. By addition, there follows the tensor character of

$$\frac{\partial}{\partial x_v} \left( \psi \frac{\partial \phi}{\partial x_\mu} \right) - \{\mu v, \tau\} \left( \psi \frac{\partial \phi}{\partial x_\tau} \right).$$

As a glance at (26) will show, this completes the demonstration for the vector

$$\psi \frac{\partial \phi}{\partial x_\mu}$$

and consequently, from what has already been proved, for any vector $A_\mu$.

By means of the extension of the vector, we may easily define the "extension" of a covariant tensor of any rank. This operation is a generalization of the extension of a vector. We restrict ourselves to the case of a tensor of the second rank, since this suffices to give a clear idea of the law of formation.

As has already been observed, any covariant tensor of the second rank can be represented [5] as the sum of tensors of the type $A_\mu B_\nu$. It will therefore be sufficient to deduce the expression for the extension of a tensor of this special type. By (26) the expressions

$$\frac{\partial A_\mu}{\partial x_\sigma} - \{\sigma\mu, \tau\} A_\tau,$$

$$\frac{\partial B_\nu}{\partial x_\sigma} - \{\sigma\nu, \tau\} B_\tau,$$

are tensors. On outer multiplication of the first by $B_\nu$ and of the second by $A_\mu$, we obtain in each case a tensor of the third rank. By adding these, we have the tensor of the third rank

$$A_{\mu\nu\sigma} = \frac{\partial B_{\mu\nu}}{\partial x_\sigma} - \{\sigma\mu, \tau\} A_{\tau\nu} - \{\sigma\nu, \tau\} A_{\mu\tau} \tag{27}$$

where we have put $A_{\mu\nu} = A_\mu B_\nu$ As the right-hand side of (27) is linear and homogeneous in the $A_{\mu\nu}$ and their first derivatives, this law of formation leads to a tensor, not only in the case of a tensor of the type $A_\mu B_\nu$ but also in the case of a sum of such tensors, i.e. in the case of any covariant tensor of the second rank. We call $A_{\mu\nu\sigma}$ the extension of the tensor $A_{\mu\nu}$.

It is clear that (26) and (24) concern only special cases of extension (the extension of the tensors of rank one and zero respectively).

In general, all special laws of formation of tensors are included in (27) in combination with the multiplication of tensors.

## § 11. Some Cases of Special Importance

*The Fundamental Tensor.*—We will first prove some lemmas which will be useful hereafter. By the rule for the differentiation of determinants

$$dg = g^{\mu\nu} g\, dg_{\mu\nu} = -g_{\mu\nu} g\, dg^{\mu\nu} \tag{28}$$

The last member is obtained from the last but one, if we bear in mind that $g_{\mu\nu} g^{\mu'\nu} = \delta_\mu^{\mu'}$, so that $g_{\mu\nu} g^{\mu\nu} = 4$, and consequently

$$g_{\mu\nu}\, dg^{\mu\nu} + g^{\mu\nu}\, dg_{\mu\nu} = 0.$$

---

5    By outer multiplication of the vector with arbitrary components $A_{11}$, $A_{12}$, $A_{13}$, $A_{14}$ by the vector with components $1, 0, 0, 0$, we produce a tensor with components

$$\begin{matrix} A_{11} & A_{12} & A_{13} & A_{14} \\ 0 & 0 & 0 & 0 \\ 0 & 0 & 0 & 0 \\ 0 & 0 & 0 & 0. \end{matrix}$$

By the addition of four tensors of this type, we obtain the tensor $A_{\mu\nu}$ with any assigned components.

From (28), it follows that

$$\frac{1}{\sqrt{-g}}\frac{\partial\sqrt{-g}}{\partial x_\sigma} = \frac{1}{2}\frac{\partial\log(-g)}{\partial x_\sigma} = \frac{1}{2}g^{\mu\nu}\frac{\partial g_{\mu\nu}}{\partial x_\sigma} = \frac{1}{2}g_{\mu\nu}\frac{\partial g^{\mu\nu}}{\partial x_\sigma}. \tag{29}$$

Further, from $g_{\mu\sigma}g^{\nu\sigma} = \delta_\mu^\nu$ it follows on differentiation that

$$\left.\begin{aligned} g_{\mu\sigma}dg^{\nu\sigma} &= -g^{\nu\sigma}dg_{\mu\sigma} \\ g_{\mu\sigma}\frac{\partial g^{\nu\sigma}}{\partial x_\lambda} &= -g^{\nu\sigma}\frac{\partial g_{\mu\sigma}}{\partial x_\lambda} \end{aligned}\right\} \tag{30}$$

From these, by mixed multiplication by $g^{\sigma\tau}$ and $g_{\nu\lambda}$ respectively, and a change of notation for the indices, we have

$$\left.\begin{aligned} dg^{\mu\nu} &= -g^{\mu\alpha}g^{\nu\beta}dg_{\alpha\beta} \\ \frac{\partial g^{\mu\nu}}{\partial x_\sigma} &= -g^{\mu\alpha}g^{\nu\beta}\frac{\partial g_{\alpha\beta}}{x_\sigma} \end{aligned}\right\} \tag{31}$$

and

$$\left.\begin{aligned} dg^{\mu\nu} &= -g_{\mu\alpha}g_{\nu\beta}dg^{\alpha\beta} \\ \frac{\partial g_{\mu\nu}}{\partial x_\sigma} &= -g_{\mu\alpha}g_{\nu\beta}\frac{\partial g^{\alpha\beta}}{\partial x_\sigma} \end{aligned}\right\} \tag{32}$$

The relation (31) admits of a transformation, of which we also have frequently to make use: From (21)

$$\frac{\partial g_{\alpha\beta}}{\partial x_\sigma} = [\alpha\sigma,\beta] + [\beta\sigma,\alpha] \tag{33}$$

Inserting this in the second formula of (31), we obtain, in view of (23)

$$\frac{\partial g^{\mu\nu}}{\partial x_\sigma} = -g^{\mu\tau}\{\tau\sigma,\nu\} - g^{\nu\tau}\{\tau\sigma,\mu\} \tag{34}$$

Substituting the right-hand side of (34) in (29), we have

$$\frac{1}{\sqrt{-g}}\frac{\partial\sqrt{-g}}{\partial x_\sigma} = \{\mu\sigma,\mu\} \tag{29a}$$

*The "Divergence" of a Contravariant Vector.*—If we take the inner product of (26) by the contravariant fundamental tensor $g^{\mu\nu}$, the right-hand side, after a transformation of the first term, assumes the form

$$\frac{\partial}{\partial x_v}\left(g^{\mu v}A_\mu\right) - A_\mu \frac{\partial g^{\mu v}}{\partial x_v} - \tfrac{1}{2}g^{\tau\alpha}\left(\frac{\partial g_{\mu\alpha}}{\partial x_v} + \frac{\partial g_{v\alpha}}{\partial x_\mu} - \frac{\partial g_{\mu v}}{\partial x_\alpha}\right)g^{\mu v}A_\tau.$$

In accordance with (31) and (29), the last term of this expression may be written

$$\tfrac{1}{2}\frac{\partial g^{\tau v}}{\partial x_v}A_\tau + \tfrac{1}{2}\frac{\partial g^{\tau\mu}}{\partial x_\mu}A_\tau + \frac{1}{\sqrt{-g}}\frac{\partial\sqrt{-g}}{\partial x_\alpha}g^{\mu v}A_\tau.$$

As the symbols of the indices of summation are immaterial, the first two terms of this expression cancel the second of the one above. If we then write $g^{\mu v}A_\mu = A^v$, so that $A^v$ like $A_\mu$ is an arbitrary vector, we finally obtain

$$\Phi = \frac{1}{\sqrt{-g}}\frac{\partial}{\partial x_v}\left(\sqrt{-g}\,A^v\right) \tag{35}$$

This scalar is the *divergence* of the contravariant, vector $A^v$.

*The "Curl" of a Covariant Vector.*—The second term in (26) is symmetrical in the indices $\mu$ and $v$. Therefore $A_{\mu v} - A_{v\mu}$ is a particularly simply constructed antisymmetrical tensor. We obtain

$$B_{\mu v} = \frac{\partial A_\mu}{\partial x_v} - \frac{\partial A_v}{\partial x_\mu} \tag{36}$$

*Antisymmetrical Extension of a Six-vector.*—Applying (27) to an antisymmetrical tensor of the second rank $A_{\mu v}$ forming in addition the two equations which arise through cyclic permutations of the indices, and adding these three equations, we obtain the tensor of the third rank

$$B_{\mu v\sigma} = A_{\mu v\sigma} + A_{v\sigma\mu} + A_{\sigma\mu v} + \frac{\partial A_{\mu v}}{\partial x_\sigma} + \frac{\partial A_{v\sigma}}{\partial x_\mu} + \frac{\partial A_{\sigma\mu}}{\partial x_v} \tag{37}$$

which it is easy to prove is antisymmetrical.

*The Divergence of a Six-vector.*—Taking the mixed product of (27) by $g^{\mu\alpha}g^{v\beta}$, we also obtain a tensor. The first term on the right-hand side of (27) may be written in the form

$$\frac{\partial}{\partial x_\sigma}\left(g^{\mu\alpha}g^{v\beta}A_{\mu v}\right) - g^{\mu\alpha}\frac{\partial g^{v\beta}}{\partial x_\sigma}A_{\mu v} - g^{v\beta}\frac{\partial g^{\mu\alpha}}{\partial x_\sigma}A_{\mu v}.$$

If we write $A_\sigma^{\alpha\beta}$ for $g^{\mu\alpha}g^{v\beta}A_{\mu v\sigma}$ and $A^{\alpha\beta}$ for $g^{\mu\alpha}g^{v\beta}A_{\mu v}$, and in the transformed first term replace

$$\frac{\partial g^{v\beta}}{\partial x_\sigma} \quad \text{and} \quad \frac{\partial g^{\mu\alpha}}{\partial x_\sigma}$$

by their values as given by (34), there results from the right-hand side of (27) an expression consisting of seven terms, of which four cancel, and there remains

$$A_\sigma^{\alpha\beta} = \frac{\partial A^{\alpha\beta}}{\partial x_\sigma} + \{\sigma\gamma,\alpha\} A^{\gamma\beta} + \{\sigma\gamma,\beta\} A^{\alpha\gamma} \tag{38}$$

This is the expression for the extension of a contravariant tensor of the second rank, and corresponding expressions for the extension of contravariant tensors of higher and lower rank may also be formed.

We note that in an analogous way we may also form the extension of a mixed tensor:—

$$A_{\mu\sigma}^\alpha = \frac{\partial A_\mu^\alpha}{\partial x_\sigma} - \{\sigma\mu,\tau\} A_\tau^\alpha + \{\sigma\tau,\alpha\} A_\mu^\tau \tag{39}$$

On contracting (38) with respect to the indices $\beta$ and $\sigma$ (inner multiplication by $\delta_\beta^\sigma$), we obtain the vector

$$A^\alpha = \frac{\partial A^{\alpha\beta}}{\partial x_\beta} + \{\beta\gamma,\beta\} A^{\alpha\gamma} + \{\beta\gamma,\alpha\} A^{\gamma\beta}.$$

On account of the symmetry of $\{\beta\gamma, \alpha\}$ with respect to the indices $\beta$ and $\gamma$, the third term on the right-hand side vanishes, if $A^{\alpha\beta}$ is, as we will assume, an antisymmetrical tensor. The second term allows itself to be transformed in accordance with (29a). Thus we obtain

$$A^\alpha = \frac{1}{\sqrt{-g}} \frac{\partial\left(\sqrt{-g}\, A^{\alpha\beta}\right)}{\partial x_\beta} \tag{40}$$

This is the expression for the divergence of a contravariant six-vector.

*The Divergence of a Mixed Tensor of the Second Rank.*—Contracting (39) with respect to the indices $\alpha$ and $\sigma$, and taking (29a) into consideration, we obtain

$$\sqrt{-g}\, A_\mu = \frac{\partial\left(\sqrt{-g}\, A_\mu^\sigma\right)}{\partial x_\sigma} - \{\sigma\mu,\tau\}\sqrt{-g}\, A_\tau^\sigma \tag{41}$$

If we introduce the contravariant tensor $A^{\rho\sigma} = g^{\rho\tau} A_\tau^\sigma$ in the last term, it assumes the form

$$- [\sigma\mu,\rho]\sqrt{-g}\, A^{\rho\sigma}.$$

If, further, the tensor $A^{\rho\sigma}$ is symmetrical, this reduces to

$$-\tfrac{1}{2}\sqrt{-g}\,\frac{\partial g_{\rho\sigma}}{\partial x_\mu} A^{\rho\sigma}.$$

Had we introduced, instead of $A^{\rho\sigma}$, the covariant tensor $A_{\rho\sigma} = g_{\rho\alpha} g_{\sigma\beta} A^{\alpha\beta}$, which is also symmetrical, the last term, by virtue of (31), would assume the form

$$\tfrac{1}{2}\sqrt{-g}\,\frac{\partial g^{\rho\sigma}}{\partial x_\mu} A_{\rho\sigma}.$$

In the case of symmetry in question, (41) may therefore be replaced by the two forms

$$\sqrt{-g}\,A_\mu = \frac{\partial\left(\sqrt{-g}\,A_\mu^\sigma\right)}{\partial x_\sigma} - \tfrac{1}{2}\frac{\partial g_{\rho\sigma}}{\partial x_\mu}\sqrt{-g}\,A^{\rho\sigma} \tag{41a}$$

$$\sqrt{-g}\,A_\mu = \frac{\partial\left(\sqrt{-g}\,A_\mu^\sigma\right)}{\partial x_\sigma} - \tfrac{1}{2}\frac{\partial g^{\rho\sigma}}{\partial x_\mu}\sqrt{-g}\,A_{\rho\sigma} \tag{41b}$$

which we have to employ later on.

## § 12. The Riemann-Christoffel Tensor

We now seek the tensor which can be obtained from the fundamental tensor *alone,* by dif-ferentiation. At first sight the solution seems obvious. We place the fundamental tensor of the $g_{\mu\nu}$ in (27) instead of any given tensor $A_{\mu\nu}$, and thus have a new tensor, namely, the extension of the fundamental tensor. But we easily convince ourselves that this extension vanishes identically. We reach our goal, however, in the following way. In (27) place

$$A_{\mu\nu} = \frac{\partial A_\mu}{\partial x_\nu} - \{\mu\nu,\rho\}\,A_\rho,$$

i.e. the extension of the four-vector $A_\mu$. Then (with a somewhat different naming of the indices) we get the tensor of the third rank

$$A_{\mu\sigma\tau} = \frac{\partial^2 A_\mu}{\partial x_\sigma \partial x_\tau} - \{\mu\sigma,\rho\}\frac{\partial A_\rho}{\partial x_\tau} - \{\mu\tau,\rho\}\frac{\partial A_\rho}{\partial x_\sigma} - \{\sigma\tau,\rho\}\frac{\partial A_\mu}{\partial x_\rho}$$
$$+ \left[-\frac{\partial}{\partial x_\tau}\{\mu\sigma,\rho\} + \{\mu\tau,\alpha\}\{\alpha\sigma,\rho\} + \{\sigma\tau,\alpha\}\{\alpha\mu,\rho\}\right]A_\rho.$$

This expression suggests forming the tensor $A_{\mu\sigma\tau} - A_{\mu\tau\sigma}$. For, if we do so, the following terms of the expression for $A_{\mu\sigma\tau}$ cancel those of $A_{\mu\tau\sigma}$, the first, the fourth, and the member corresponding to the last term in square brackets; because all these are symmetrical in $\sigma$ and $\tau$. The same holds good for the sum of the second and third terms. Thus we obtain

$$A_{\mu\sigma\tau} - A_{\mu\tau\sigma} = B^\rho_{\mu\sigma\tau}A_\rho \tag{42}$$

where

$$B^\rho_{\mu\sigma\tau} = \frac{\partial}{\partial x_\tau}\{\mu\sigma,\rho\} + \frac{\partial}{\partial x_\sigma}\{\mu\tau,\rho\} - \{\mu\sigma,\alpha\}\{\alpha\tau,\rho\} + \{\mu\tau,\alpha\}\{\alpha\sigma,\rho\} \tag{43}$$

The essential feature of the result is that on the right side of (42) the $A_\rho$ occur alone, without their derivatives. From the tensor character of $A_{\mu\sigma\tau} - A_{\mu\tau\sigma}$ in conjunction with the fact that $A_\rho$ is an arbitrary vector, it follows, by reason of § 7, that $B^\rho_{\mu\sigma\tau}$ is a tensor (the Riemann-Christoffel tensor).

The mathematical importance of this tensor is as follows: If the continuum is of such a nature that there is a co-ordinate system with reference to which the $g_{\mu\nu}$ constants, then all the $B^\rho_{\mu\sigma\tau}$ vanish. If we choose any new system of co-ordinates in place of the original ones, the $g_{\mu\nu}$ referred thereto will not be constants, but in consequence of its tensor nature, the transformed components of $B^\rho_{\mu\sigma\tau}$ will still vanish in the new system. Thus the vanishing of the Riemann tensor is a necessary condition that, by an appropriate choice of the system of reference, the $g_{\mu\nu}$ may be constants. In our problem this corresponds to the case in which,[6] with a suitable choice of the system of reference, the special theory of relativity holds good for a *finite* region of the continuum.

Contracting (43) with respect to the indices $\tau$ and $\rho$ we obtain the covariant tensor of second rank

where

$$\left.\begin{aligned}
G_{\mu\nu} &= B^\rho_{\mu\nu\rho} = R_{\mu\nu} + S_{\mu\nu} \\[2mm]
R_{\mu\nu} &= -\frac{\partial}{\partial x_\alpha}\{\mu\nu,\alpha\} + \{\mu\alpha,\beta\}\{\nu\beta,\alpha\} \\[2mm]
S_{\mu\nu} &= \frac{\partial^2 \log\sqrt{-g}}{\partial x_\mu \partial x_\nu} - \{\mu\nu,\alpha\}\frac{\partial \log\sqrt{-g}}{\partial x_\alpha}
\end{aligned}\right\} \tag{44}$$

*Note on the Choice of Co-ordinates.*—It has already been observed in § 8, in connection with equation (18a), that the choice of co-ordinates may with advantage be made so that $\sqrt{-g} = 1$. A glance at the equations obtained in the last two sections shows that by such a choice the laws of formation of tensors undergo an important simplification. This applies particularly to $G_{\mu\nu}$, the tensor just developed, which plays a fundamental part in the theory to be set forth. For this specialization of the choice of co-ordinates brings about the vanishing of $S_{\mu\nu}$, so that the tensor $G_{\mu\nu}$ reduces to $R_{\mu\nu}$.

On this account I shall hereafter give all relations in the simplified form which this specialization of the choice of coordinates brings with it. It will then be an easy matter to revert to the *generally* covariant equations, if this seems desirable in a special case.

## C. THEORY OF THE GRAVITATIONAL FIELD

### § 13. Equations of Motion of a Material Point in the Gravitational Field. Expression for the Field-components of Gravitation

A freely movable body not subjected to external forces moves, according to the special theory of relativity, in a straight line and uniformly. This is also the case, according to the general theory of relativity, for a part of four-dimensional space in which the system of co-ordinates $K_0$, may be, and is, so chosen that they have the special constant values given in (4).

---

6    The mathematicians have proved that this is also a *sufficient* condition.

If we consider precisely this movement from any chosen system of co-ordinates $K_1$, the body, observed from $K_1$, moves, according to the considerations in § 2, in a gravitational field. The law of motion with respect to $K_1$ results without difficulty from the following consideration. With respect to $K_0$ the law of motion corresponds to a four-dimensional straight line, i.e. to a geodetic line. Now since the geodetic line is defined independently of the system of reference, its equations will also be the equation of motion of the material point with respect to $K_1$. If we set

$$\Gamma^{\tau}_{\mu\nu} = -\{\mu\nu, \tau\} \tag{45}$$

the equation of the motion of the point with respect to $K_1$ becomes

$$\frac{d^2 x_{\tau}}{ds^2} = \Gamma^{\tau}_{\mu\nu} \frac{dx_{\mu}}{ds} \frac{dx_{\nu}}{ds} \tag{46}$$

We now make the assumption, which readily suggests itself, that this covariant system of equations also defines the motion of the point in the gravitational field in the case when there is no system of reference $K_0$, with respect to which the special theory of relativity holds good in a finite region. We have all the more justification for this assumption as (46) contains only *first* derivatives of the $g_{\mu\nu}$, between which even in the special case of the existence of $K_0$, no relations subsist.[7]

If the $\Gamma^{\tau}_{\mu\nu}$ vanish, then the point moves uniformly in a straight line. These quantities therefore condition the deviation of the motion from uniformity. They are the components of the gravitational field.

## § 14. The Field Equations of Gravitation in the Absence of Matter

We make a distinction hereafter between "gravitational field" and "matter" in this way, that we denote everything but the gravitational field as "matter." Our use of the word therefore includes not only matter in the ordinary sense, but the electromagnetic field as well.

Our next task is to find the field equations of gravitation in the absence of matter. Here we again apply the method employed in the preceding paragraph in formulating the equations of motion of the material point. A special case in which the required equations must in any case be satisfied is that of the special theory of relativity, in which the $g_{\mu\nu}$ have certain constant values. Let this be the case in a certain finite space in relation to a definite system of co-ordinates $K_0$. Relatively to this system all the components of the Riemann tensor $B^{\rho}_{\mu\sigma\tau}$, defined in (43), vanish. For the space under consideration they then vanish, also in any other system of co-ordinates.

Thus the required equations of the matter-free gravitational field must in any case be satisfied if all $B^{\rho}_{\mu\sigma\tau}$ vanish. But this condition goes too far. For it is clear that, e.g., the gravitational field generated by a material point in its environment certainly cannot be "transformed away" by any choice of the system of co-ordinates, i.e. it cannot be transformed to the case of constant $g_{\mu\nu}$.

---

7    It is only between the second (and first) derivatives that, by § 12, the relations $B^{\rho}_{\mu\sigma\tau} = 0$ subsist.

This prompts us to require for the matter-free gravitational field that the symmetrical tensor $G_{\mu\nu}$, derived from the tensor $B^{\rho}_{\mu\nu\tau}$, shall vanish. Thus we obtain ten equations for the ten quantities $g_{\mu\nu}$, which are satisfied in the special case of the vanishing of all $B^{\rho}_{\mu\nu\tau}$. With the choice which we have made of a system of co-ordinates, and taking (44) into consideration, the equations for the matter-free field are

$$\left.\begin{aligned} \frac{\partial \Gamma^{\alpha}_{\mu\nu}}{\partial x_{\alpha}} + \Gamma^{\alpha}_{\mu\beta}\Gamma^{\beta}_{\nu\alpha} &= 0 \\ \sqrt{-g} &= 1 \end{aligned}\right\} \tag{47}$$

It must be pointed out that there is only a minimum of arbitrariness in the choice of these equations. For besides $G_{\mu\nu}$ there is no tensor of second rank which is formed from the $g_{\mu\nu}$ and its derivatives, contains no derivations higher than second, and is linear in these derivatives.[8]

These equations, which proceed, by the method of pure mathematics, from the requirement of the general theory of relativity, give us, in combination with the equations of motion (46), to a first approximation Newton's law of attraction, and to a second approximation the explanation of the motion of the perihelion of the planet Mercury discovered by Leverrier (as it remains after corrections for perturbation have been made). These facts must, in my opinion, be taken as a convincing proof of the correctness of the theory.

### § 15. The Hamiltonian Function for the Gravitational Field. Laws of Momentum and Energy

To show that the field equations correspond to the laws of momentum and energy, it is most convenient to write them in the following Hamiltonian form:—

$$\left.\begin{aligned} \delta \int H d\tau &= 0 \\ H &= g^{\mu\nu}\Gamma^{\alpha}_{\mu\beta}\Gamma^{\beta}_{\nu\alpha} \\ \sqrt{-g} &= 1 \end{aligned}\right\} \tag{47a}$$

where, on the boundary of the finite four-dimensional region of integration which we have in view, the variations vanish.

We first have to show that the form (47a) is equivalent to the equations (47). For this purpose we regard H as a function of the $g^{\mu\nu}$ and the $g^{\mu\nu}_{\sigma}(=\partial g^{\mu\nu}/\partial x_{\sigma})$.

Then in the first place

---

8    Properly speaking, this can be affirmed only of the tensor

$$G_{\mu\nu} + \lambda g_{\mu\nu}g^{\alpha\beta}G_{\alpha\beta},$$

where $\lambda$ is a constant. If, however, we set this tensor $= 0$, we come back again to the equations $G_{\mu\nu} = 0$.

$$\delta H = \Gamma^{\alpha}_{\mu\beta}\Gamma^{\beta}_{\nu\alpha}\delta g^{\mu\nu} + 2g^{\mu\nu}\Gamma^{\alpha}_{\mu\beta}\delta\Gamma^{\beta}_{\nu\alpha}$$
$$= -\Gamma^{\alpha}_{\mu\beta}\Gamma^{\beta}_{\nu\alpha}\delta g^{\mu\nu} + 2\Gamma^{\alpha}_{\mu\beta}\delta\left(g^{\mu\nu}\Gamma^{\beta}_{\nu\alpha}\right).$$

But

$$\delta\left(g^{\mu\nu}\Gamma^{\beta}_{\nu\alpha}\right) = -\tfrac{1}{2}\delta\left[g^{\mu\nu}g^{\beta\lambda}\left(\frac{\partial g_{\nu\lambda}}{\partial x_{\alpha}} + \frac{\partial g_{\alpha\lambda}}{\partial x_{\nu}} - \frac{\partial g_{\alpha\nu}}{\partial x_{\lambda}}\right)\right].$$

The terms arising from the last two terms in round brackets are of different sign, and result from each other (since the denomination of the summation indices is immaterial) through interchange of the indices $\mu$ and $\beta$. They cancel each other in the expression for $\delta H$, because they are multiplied by the quantity $\Gamma^{\alpha}_{\mu\beta}$, which is symmetrical with respect to the indices $\mu$ and $\beta$. Thus there remains only the first term in round brackets to be considered, so that, taking (31) into account, we obtain

$$\delta H = -\Gamma^{\alpha}_{\mu\beta}\Gamma^{\beta}_{\nu\alpha}\delta g^{\mu\nu} + \Gamma^{\alpha}_{\mu\beta}\delta g^{\mu\beta}_{\alpha}.$$

Thus

$$\left.\begin{array}{c}\dfrac{\partial H}{\partial g^{\mu\nu}} = -\Gamma^{\alpha}_{\mu\beta}\Gamma^{\beta}_{\nu\alpha} \\[2mm] \dfrac{\partial H}{\partial g^{\mu\nu}_{\sigma}} = \Gamma^{\sigma}_{\mu\nu}\end{array}\right\} \tag{48}$$

Carrying out the variation in (47a), we get in the first place

$$\frac{\partial}{\partial x_{\alpha}}\left(\frac{\partial H}{\partial g^{\mu\nu}_{\alpha}}\right) - \frac{\partial H}{\partial g^{\mu\nu}} = 0, \tag{47b}$$

which, on account of (48), agrees with (47), as was to be proved.
    If we multiply (47b) by $g^{\mu\nu}_{\sigma}$, then because

$$\frac{\partial g^{\mu\nu}_{\sigma}}{\partial x_{\alpha}} = \frac{\partial g^{\mu\nu}_{\alpha}}{\partial x_{\sigma}}$$

and, consequently,

$$g^{\mu\nu}_{\sigma}\frac{\partial}{\partial x_{\alpha}}\left(\frac{\partial H}{\partial g^{\mu\nu}_{\alpha}}\right) = \frac{\partial}{\partial x_{\alpha}}\left(g^{\mu\nu}_{\sigma}\frac{\partial H}{\partial g^{\mu\nu}_{\alpha}}\right) - \frac{\partial H}{\partial g^{\mu\nu}_{\alpha}}\frac{\partial g^{\mu\nu}_{\alpha}}{\partial x_{\sigma}},$$

we obtain the equation

$$\frac{\partial}{\partial x_{\alpha}}\left(g^{\mu\nu}_{\sigma}\frac{\partial H}{\partial g^{\mu\nu}_{\alpha}}\right) - \frac{\partial H}{\partial x_{\sigma}} = 0$$

or[9]

$$
\left.\begin{aligned}
\frac{\partial t_\sigma^\alpha}{\partial x_\alpha} &= 0 \\[2mm]
-2\kappa t_\sigma^\alpha &= g_\sigma^{\mu\nu}\frac{\partial H}{\partial g_\alpha^{\mu\nu}} - \delta_\sigma^\alpha H
\end{aligned}\right\}
\tag{49}
$$

where, on account of (48), the second equation of (47), and (34)

$$
\kappa t_\sigma^\alpha = \tfrac{1}{2}\delta_\sigma^\alpha g^{\mu\nu}\Gamma_{\mu\beta}^\lambda \Gamma_{\nu\lambda}^\beta - g^{\mu\nu}\Gamma_{\mu\beta}^\alpha \Gamma_{\nu\sigma}^\beta
\tag{50}
$$

It is to be noticed that $t_\sigma^\alpha$ is not a tensor; on the other hand (49) applies to all systems of co-ordinates for which $\sqrt{-g} = 1$. This equation expresses the law of conservation of momentum and of energy for the gravitational field. Actually the integration of this equation over a three-dimensional volume V yields the four equations

$$
\frac{d}{dx_4}\int t_\sigma^4 dV = \int \left(lt_\sigma^1 + mt_\sigma^2 + nt_\sigma^3\right)ds
\tag{49a}
$$

where $l, m, n$ denote the direction-cosines of direction of the inward drawn normal at the element $dS$ of the bounding surface (in the sense of Euclidean geometry). We recognize in this the expression of the laws of conservation in their usual form. The quantities $t_\sigma^\alpha$ we call the "energy components" of the gravitational field.

I will now give equations (47) in a third form, which is particularly useful for a vivid grasp of our subject. By multiplication of the field equations (47) by $g^{\nu\sigma}$ these are obtained in the "mixed" form. Note that

$$
g^{\nu\sigma}\frac{\partial \Gamma_{\mu\nu}^\alpha}{\partial x_\alpha} = \frac{\partial}{\partial x_\alpha}\left(g^{\nu\sigma}\Gamma_{\mu\nu}^\alpha\right) - \frac{\partial g^{\nu\sigma}}{\partial x_\alpha}\Gamma_{\mu\nu}^\alpha,
$$

which quantity, by reason of (34), is equal to

$$
\frac{\partial}{\partial x_\alpha}\left(g^{\nu\sigma}\Gamma_{\mu\nu}^\alpha\right) - g^{\nu\beta}\Gamma_{\alpha\beta}^\sigma\Gamma_{\mu\nu}^\alpha - g^{\sigma\beta}\Gamma_{\beta\alpha}^\nu\Gamma_{\mu\nu}^\alpha,
$$

or (with different symbols for the summation indices)

$$
\frac{\partial}{\partial x_\alpha}\left(g^{\sigma\beta}\Gamma_{\mu\beta}^\alpha\right) - g^{\gamma\delta}\Gamma_{\gamma\beta}^\sigma\Gamma_{\delta\mu}^\beta - g^{\nu\sigma}\Gamma_{\mu\beta}^\alpha\Gamma_{\nu\alpha}^\beta.
$$

The third term of this expression cancels with the one arising from the second term of the field equations (47); using relation (50), the second term may be written

---

9   The reason for the introduction of the factor $-2\kappa$ will be apparent later.

$$\kappa\left(t_\mu^\sigma - \tfrac{1}{2}\delta_\mu^\sigma t\right),$$

where $t = t_\alpha^\alpha$. Thus instead of equations (47) we obtain

$$\left.\begin{array}{c} \dfrac{\partial}{\partial x_\alpha}\left(g^{\alpha\beta}\Gamma_{\mu\beta}^\alpha\right) = -\kappa\left(t_\mu^\sigma - \tfrac{1}{2}\delta_\mu^\sigma t\right) \\[2ex] \sqrt{-g} = 1 \end{array}\right\} \tag{51}$$

## § 16. The General Form of the Field Equations of Gravitation

The field equations for matter-free space formulated in § 15 are to be compared with the field equation

$$\nabla^2\phi = 0$$

of Newton's theory. We require the equation corresponding to Poisson's equation

$$\nabla^2\phi = 4\pi\kappa\rho,$$

where $\rho$ denotes the density of matter.

The special theory of relativity has led to the conclusion that inert mass is nothing more or less than energy, which finds its complete mathematical expression in a symmetrical tensor of second rank, the energy-tensor. Thus in the general theory of relativity we must introduce a corresponding energy-tensor of matter $T_\sigma^\alpha$, which, like the energy-components $t_\sigma$ [equations (49) and (50)] of the gravitational field, will have mixed character, but will pertain to a symmetrical covariant tensor.[10]

The system of equation (51) shows how this energy-tensor (corresponding to the density $\rho$ in Poisson's equation) is to be introduced into the field equations of gravitation. For if we consider a complete system (e.g. the solar system), the total mass of the system, and therefore its total gravitating action as well, will depend on the total energy of the system, and therefore on the ponderable energy together with the gravitational energy. This will allow itself to be expressed by introducing into (51), in place of the energy-components of the gravitational field alone, the sums $t_\mu^\sigma + T_\mu^\sigma$ of the energy-components of matter and of gravitational field. Thus instead of (51) we obtain the tensor equation

$$\left.\begin{array}{c} \dfrac{\partial}{\partial x_\alpha}\left(g^{\sigma\beta}T_{\mu\beta}^\alpha\right) = -\kappa\left[\left(t_\mu^\sigma + T_\mu^\sigma\right) - \tfrac{1}{2}\delta_\mu^\sigma(t + T)\right] \\[2ex] \sqrt{-g} = 1 \end{array}\right\} \tag{52}$$

10   $g_{\alpha\tau}T_\sigma^\alpha = T_{\sigma\tau}$ and $g^{\sigma\beta}T_\sigma^\alpha = T^{\alpha\beta}$ are to be symmetrical tensors.

where we have set $T = T_\mu^\mu$ (Laue's scalar). These are the required general field equations of gravitation in mixed form. Working back from these, we have in place of (47)

$$\frac{\partial}{\partial x_\alpha} \Gamma_{\mu\nu}^\alpha + \Gamma_{\mu\beta}^\alpha \Gamma_{\nu\alpha}^\beta = -\kappa\left(T_{\mu\nu} - \tfrac{1}{2} g_{\mu\nu} T\right), \left.\begin{array}{c} \\ \\ \end{array}\right\}$$
$$\sqrt{-g} = 1 \qquad\qquad\qquad\qquad (53)$$

It must be admitted that this introduction of the energy-tensor of matter is not justified by the relativity postulate alone. For this reason we have here deduced it from the requirement that the energy of the gravitational field shall act gravitatively in the same way as any other kind of energy. But the strongest reason for the choice of these equations lies in their consequence, that the equations of conservation of momentum and energy, corresponding exactly to equations (49) and (49a), hold good for the components of the total energy. This will be shown in § 17.

### § 17. The Laws of Conservation in the General Case

Equation (52) may readily be transformed so that the second term on the right-hand side vanishes. Contract (52) with respect to the indices $\mu$ and $\sigma$, and after multiplying the resulting equation by $\tfrac{1}{2}\delta_\mu^\sigma$, subtract it from equation (52). This gives

$$\frac{\partial}{\partial x_\alpha}\left(g^{\sigma\beta}\Gamma_{\mu\beta}^\alpha - \tfrac{1}{2}\delta_\mu^\sigma g^{\lambda\beta}\Gamma_{\lambda\beta}^\alpha\right) = -\kappa\left(t_\mu^\sigma + T_\mu^\sigma\right). \qquad (52a)$$

On this equation we perform the operation $\partial/\partial x_\sigma$. We have

$$\frac{\partial^2}{\partial x_\alpha \partial x_\sigma}\left(g^\sigma \Gamma_{\beta\mu}^\alpha\right) = -\tfrac{1}{2}\frac{\partial^2}{\partial x_\alpha \partial x_\sigma}\left[g^{\sigma\beta} g^{\alpha\lambda}\left(\frac{\partial g_{\mu\lambda}}{\partial x_\beta} + \frac{\partial g_{\beta\lambda}}{\partial x_\mu} - \frac{\partial g_{\mu\beta}}{\partial x_\lambda}\right)\right].$$

The first and third terms of the round brackets yield contributions which cancel one another, as may be seen by interchanging, in the contribution of the third term, the summation indices $\alpha$ and $\sigma$ on the one hand, and $\beta$ and $\lambda$ on the other. The second term may be re-modelled by (31), so that we have

$$\frac{\partial^2}{\partial x_\alpha \partial x_\sigma}\left(g^{\sigma\beta}\Gamma_{\mu\beta}^\alpha\right) = \tfrac{1}{2}\frac{\partial^3 g^{\alpha\beta}}{\partial x_\alpha \partial x_\beta \partial x_\mu} \qquad (54)$$

The second term on the left-hand side of (52a) yields in the first place

$$-\tfrac{1}{2}\frac{\partial^2}{\partial x_\alpha \partial x_\mu}\left(g^{\lambda\beta}\Gamma_{\lambda\beta}^\alpha\right)$$

or

$$\frac{1}{4}\frac{\partial^2}{\partial x_\alpha \partial x_\mu}\left[g^{\lambda\beta}g^{\alpha\delta}\left(\frac{\partial g_{\delta\lambda}}{\partial x_\beta} + \frac{\partial g_{\delta\beta}}{\partial x_\lambda} - \frac{\partial g_{\lambda\beta}}{\partial x_\delta}\right)\right].$$

With the choice of co-ordinates which we have made, the term deriving from the last term in round brackets disappears by reason of (29). The other two may be combined, and together, by (31), they give

$$-\frac{1}{2}\frac{\partial^3 g^{\alpha\beta}}{\partial x_\alpha \partial x_\beta \partial x_\mu},$$

so that in consideration of (54), we have the identity

$$\frac{\partial^2}{\partial x_\alpha \partial x_\sigma}\left(g^{\rho\beta}\Gamma_{\mu\beta} - \frac{1}{2}\delta_\mu^\sigma g^{\lambda\beta}\Gamma_{\lambda\beta}^\alpha\right) \equiv 0 \tag{55}$$

From (55) and (52a), it follows that

$$\frac{\partial\left(t_\mu^\sigma + T_\mu^\sigma\right)}{\partial x_\sigma} = 0 \tag{56}$$

Thus it results from our field equations of gravitation that the laws of conservation of momentum and energy are satisfied. This may be seen most easily from the consideration which leads to equation (49a); except that here, instead of the energy components $t^\sigma$ of the gravitational field, we have to introduce the totality of the energy components of matter and gravitational field.

### § 18. The Laws of Momentum and Energy for Matter, as a Consequence of the Field Equations

Multiplying (53) by $\partial g^{\mu\nu}/\partial x_\sigma$, we obtain, by the method adopted in § 15, in view of the vanishing of

$$g_{\mu\nu}\frac{\partial g^{\mu\nu}}{\partial x_\sigma},$$

the equation

$$\frac{\partial t_\sigma^\alpha}{\partial x_\alpha} + \frac{1}{2}\frac{\partial g^{\mu\nu}}{\partial x_\sigma}T_{\mu\nu} = 0,$$

or, in view of (56),

$$\frac{\partial T_\sigma^\alpha}{\partial x_\alpha} + \frac{1}{2}\frac{\partial g^{\mu\nu}}{\partial x_\sigma}T_{\mu\nu} = 0 \tag{57}$$

Comparison with (41b) shows that with the choice of system of co-ordinates which we have made, this equation predicates nothing more or less than the vanishing of divergence of the material energy-tensor. Physically, the occurrence of the second term on the left-hand side shows that laws of conservation of momentum and energy do not apply in the strict sense for matter alone, or else that they apply only when the $g^{\mu\nu}$ are constant, i.e. when the field intensities of gravitation vanish. This second term is an expression for momentum, and for energy, as transferred per unit of volume and time from the gravitational field to matter. This is brought out still more clearly by re-writing (57) in the sense of (41) as

$$\frac{\partial \mathrm{T}_\sigma^\alpha}{\partial x_\alpha} = -\Gamma_{\alpha\sigma}^\beta \mathrm{T}_\beta^\alpha \tag{57a}$$

The right side expresses the energetic effect of the gravitational field on matter.

Thus the field equations of gravitation contain four conditions which govern the course of material phenomena. They give the equations of material phenomena completely, if the latter is capable of being characterized by four differential equations independent of one another.[11]

## D. MATERIAL PHENOMENA

The mathematical aids developed in part B enable us forthwith to generalize the physical laws of matter (hydrodynamics, Maxwell's electrodynamics), as they are formulated in the special theory of relativity, so that they will fit in with the general theory of relativity. When this is done, the general principle of relativity does not indeed afford us a further limitation of possibilities; but it makes us acquainted with the influence of the gravitational field on all processes, without our having to introduce any new hypothesis whatever.

Hence it comes about that it is not necessary to introduce definite assumptions as to the physical nature of matter (in the narrower sense). In particular it may remain an open question whether the theory of the electromagnetic field in conjunction with that of the gravitational field furnishes a sufficient basis for the theory of matter or not. The general postulate of relativity is unable on principle to tell us anything about this. It must remain to be seen, during the working out of the theory, whether electromagnetics and the doctrine of gravitation are able in collaboration to perform what the former by itself is unable to do.

### § 19. Euler's Equations for a Frictionless Adiabatic Fluid

Let $p$ and $\rho$ be two scalars, the former of which we call the "pressure," the latter the "density" of a fluid; and let an equation subsist between them. Let the contravariant symmetrical tensor

---

11  On this question cf. H. Hilbert, Nachr. d. K. Gesellsch. d. Wiss. zu Göttingen, Math.-phys. Klasse, 1915, p. 3.

$$T^{\alpha\beta} = -g^{\alpha\beta}p + \rho\frac{dx_\alpha}{ds}\frac{dx_\beta}{ds}$$

be the contravariant energy-tensor of the fluid. To it belongs the covariant tensor

$$T_{\mu\nu} = -g_{\mu\nu}p + g_{\mu\alpha}g_{\mu\beta}\frac{dx_\alpha}{ds}\frac{dx_\beta}{ds}\rho, \qquad (58a)$$

as well as the mixed tensor [12]

$$T^\alpha_\sigma = -\delta^\alpha_\sigma p + g_{\sigma\beta}\frac{dx_\beta}{ds}\frac{dx_\alpha}{ds}\rho \qquad (58b)$$

Inserting the right-hand side of (58b) in (57a), we obtain the Eulerian hydrodynamical equations of the general theory of relativity. They give, in theory, a complete solution of the problem of motion, since the four equations (57a), together with the given equation between $p$ and $\rho$, and the equation

$$g_{\alpha\beta}\frac{dx_\alpha}{ds}\frac{dx_\beta}{ds} = 1,$$

are sufficient, $g_{\alpha\beta}$ being given, to define the six unknowns

$$p,\ \rho,\ \frac{dx_1}{ds},\ \frac{dx_2}{ds},\ \frac{dx_3}{ds},\ \frac{dx_4}{ds}.$$

If the $g_{\mu\nu}$ are also unknown, the equations (53) are brought in. These are eleven equations for defining the ten functions $g_{\mu\nu}$, so that these functions appear over-defined. We must remember, however, that the equations (57a) are already contained in the equations (53), so that the latter represent only seven independent equations. There is good reason for this lack of definition, in that the wide freedom of the choice of co-ordinates causes the problem to remain mathematically undefined to such a degree that three of the functions of space may be chosen at will.[13]

## § 20. Maxwell's Electromagnetic Field Equations for Free Space

Let $\phi_\nu$ be the components of a covariant vector—the electromagnetic potential vector. From them we form, in accordance with (36), the components $F_{\rho\sigma}$ of the covariant six-vector of the electromagnetic field, in accordance with the system of equations

---

12   For an observer using a system of reference in the sense of the special theory of relativity for an infinitely small region, and moving with it, the density of energy $T_4^4$ equals $\rho - p$. This gives the definition of $\rho$. Thus $\rho$ is not constant for an incompressible fluid.

13   On the abandonment of the choice of co-ordinates with $g = -1$, there remain *four* functions of space with liberty of choice, corresponding to the four arbitrary functions at our disposal in the choice of co-ordinates.

$$F_{\rho\sigma} = \frac{\partial \phi_\rho}{\partial x_\sigma} - \frac{\partial \phi_\sigma}{\partial x_\rho} \tag{59}$$

It follows from (59) that the system of equations

$$\frac{\partial F_{\rho\sigma}}{\partial x_\tau} + \frac{\partial F_{\sigma\tau}}{\partial x_\rho} + \frac{\partial F_{\tau\rho}}{\partial x_\sigma} = 0 \tag{60}$$

is satisfied, its left side being, by (37), an antisymmetrical tensor of the third rank. System (60) thus contains essentially four equations which are written out as follows:—

$$\left.\begin{aligned}
\frac{\partial F_{23}}{\partial x_4} + \frac{\partial F_{34}}{\partial x_2} + \frac{\partial F_{42}}{\partial x_3} &= 0 \\[4pt]
\frac{\partial F_{34}}{\partial x_1} + \frac{\partial F_{41}}{\partial x_3} + \frac{\partial F_{13}}{\partial x_4} &= 0 \\[4pt]
\frac{\partial F_{41}}{\partial x_2} + \frac{\partial F_{12}}{\partial x_4} + \frac{\partial F_{24}}{\partial x_1} &= 0 \\[4pt]
\frac{\partial F_{12}}{\partial x_3} + \frac{\partial F_{23}}{\partial x_1} + \frac{\partial F_{31}}{\partial x_2} &= 0
\end{aligned}\right\} \tag{60a}$$

This system corresponds to the second of Maxwell's systems of equations. We recognize this at once by setting

$$\left.\begin{aligned}
F_{23} &= H_x, & F_{14} &= E_x \\
F_{31} &= H_y, & F_{24} &= E_y \\
F_{12} &= H_z, & F_{34} &= E_z
\end{aligned}\right\} \tag{61}$$

Then in place of (60a) we may set, in the usual notation of three-dimensional vector analysis,

$$\left.\begin{aligned}
-\frac{\partial H}{\partial t} &= \operatorname{curl} E \\[4pt]
\operatorname{div} H &= 0
\end{aligned}\right\} \tag{60b}$$

We obtain Maxwell's first system by generalizing the form given by Minkowski. We introduce the contravariant six-vector associated with $F^{\alpha\beta}$

$$F^{\mu\nu} = g^{\mu\alpha} g^{\nu\beta} F_{\alpha\beta} \tag{62}$$

and also the contravariant vector $J^\mu$ of the density of the electric current. Then, taking (40) into consideration, the following equations will be invariant for any substitution whose invariant is unity (in agreement with the chosen coordinates) :—

$$\frac{\partial}{\partial x_\nu} F^{\mu\nu} = J^\mu \tag{63}$$

Let

$$
\left.\begin{aligned}
F^{23} &= H'_x, \quad F^{14} = -E'_x \\
F^{31} &= H'_y, \quad F^{24} = -E'_y \\
F^{12} &= H'_z, \quad F^{34} = -E'_z
\end{aligned}\right\}
\tag{64}
$$

which quantities are equal to the quantities $H_x \ldots E^z$ in the special case of the restricted theory of relativity; and in addition

$$
J^1 = j_x, J^2 = j_y, J^3 = j_z, J^4 = \rho,
$$

we obtain in place of (63)

$$
\left.\begin{aligned}
\frac{\partial E'}{\partial t} + j &= \text{curl } H' \\
\text{div } E' &= \rho
\end{aligned}\right\}
\tag{63a}
$$

The equations (60), (62), and (63) thus form the generalization of Maxwell's field equations for free space, with the convention which we have established with respect to the choice of co-ordinates.

*The Energy-components of the Electromagnetic Field.*—We form the inner product

$$
\kappa_\sigma = F_{\sigma\mu} J^\mu
\tag{65}
$$

By (61) its components, written in the three-dimensional manner, are

$$
\left.\begin{aligned}
\kappa_1 &= \rho E_x + [j \cdot H]^x \\
&\cdot \quad \cdot \quad \cdot \quad \cdot \\
&\cdot \quad \cdot \quad \cdot \quad \cdot \\
\kappa_4 &= -(jE)
\end{aligned}\right\}
\tag{65a}
$$

$k_\sigma$ is a covariant vector the components of which are equal to the negative momentum, or, respectively, the energy, which is transferred from the electric masses to the electromagnetic field per unit of time and volume. If the electric masses are free, that is, under the sole influence of the electromagnetic field, the covariant vector $k_\sigma$ will vanish.

To obtain the energy-components $T_\sigma^\nu$ of the electromagnetic field, we need only give to equation $k_\sigma = 0$ the form of equation (57). From (63) and (65) we have in the first place

$$
\kappa_\sigma = F_{\sigma\mu} \frac{\partial F^{\mu\nu}}{\partial x_\nu} = \frac{\partial}{\partial x_\nu}(F_{\sigma\mu} F^{\mu\nu}) - F^{\mu\rho} \frac{\partial F_{\sigma\mu}}{\partial x_\nu}.
$$

The second term of the right-hand side, by reason of (60), permits the transformation

$$F^{\mu\nu}\frac{\partial F_{\sigma\mu}}{\partial x_\nu} = -\tfrac{1}{2}F^{\mu\nu}\frac{\partial F_{\mu\nu}}{\partial x_\sigma} = -\tfrac{1}{2}g^{\mu\alpha}g^{\nu\beta}F_{\alpha\beta}\frac{\partial F_{\mu\nu}}{\partial x_\sigma},$$

which latter expression may, for reasons of symmetry, also be written in the form

$$-\tfrac{1}{4}\left[g^{\mu\alpha}g^{\nu\beta}F_{\alpha\beta}\frac{\partial F_{\mu\nu}}{\partial x_\sigma} + g^{\mu\alpha}g^{\nu\beta}\frac{\partial F_{\alpha\beta}}{\partial x_\sigma}F_{\mu\nu}\right].$$

But for this we may set

$$-\tfrac{1}{4}\frac{\partial}{\partial x_\sigma}\left(g^{\mu\alpha}g^{\nu\beta}F_{\alpha\beta}F_{\mu\nu}\right) + \tfrac{1}{4}F_{\alpha\beta}F_{\mu\nu}\frac{\partial}{\partial x_\sigma}\left(g^{\mu\alpha}g^{\nu\beta}\right).$$

The first of these terms is written more briefly

$$-\tfrac{1}{4}\frac{\partial}{\partial x_\sigma}\left(F^{\mu\nu}F_{\mu\nu}\right);$$

the second, after the differentiation is carried out, and after some reduction, results in

$$-\tfrac{1}{2}F^{\mu\tau}F_{\mu\nu}g^{\nu\rho}\frac{\partial g_{\sigma\tau}}{\partial x_\sigma}.$$

Taking all three terms together we obtain the relation

$$\kappa_\sigma = \frac{\partial T^\nu_\sigma}{\partial x_\nu} - \tfrac{1}{2}g^{\tau\mu}\frac{\partial g_{\mu\nu}}{\partial x_\sigma}T^\nu_\tau \tag{66}$$

where

$$T^\nu_\sigma = -F_{\sigma\alpha}F^{\nu\alpha} + \tfrac{1}{4}\delta^\nu_\sigma F_{\alpha\beta}F^{\alpha\beta}.$$

Equation (66), if $k_\sigma$ vanishes, is, on account of (30), equivalent to (57) or (57a) respectively. Therefore the $T^\nu_\sigma$ are the energy-components of the electromagnetic field. With the help of (61) and (64), it is easy to show that these energy-components of the electromagnetic field in the case of the special theory of relativity give the well-known Maxwell-Poynting expressions.

We have now deduced the general laws which are satisfied by the gravitational field and matter, by consistently using a system of co-ordinates for which $\sqrt{-g} = 1$. We have thereby achieved a considerable simplification of formulae and calculations, without failing to comply with the requirement of general covariance; for we have drawn our equations from generally covariant equations by specializing the system of co-ordinates.

Still the question is not without a formal interest, whether with a correspondingly generalized definition of the energy-components of gravitational field and matter, even without specializing the system of co-ordinates, it is possible to formulate laws of conservation in the form of equation (56), and field equations of gravitation of the same nature as (52) or (52a), in such a manner that on the left we have a divergence (in the ordinary sense), and on the right the sum of the energy-components of matter and gravitation. I have found that in both cases this is actually so. But I do not think that the communication of my somewhat extensive reflections on this subject would be worth while, because after all they do not give us anything that is materially new.

E

### § 21. Newton's Theory as a First Approximation

As has already been mentioned more than once, the special theory of relativity as a special case of the general theory is characterized by the $g_{\mu\nu}$ having the constant values (4). From what has already been said, this means complete neglect of the effects of gravitation. We arrive at a closer approximation to reality by considering the case where the $g_{\mu\nu}$ differ from the values of (4) by quantities which are small compared with 1, and neglecting small quantities of second and higher order. (First point of view of approximation.)

It is further to be assumed that in the space-time territory under consideration the $g_{\mu\nu}$ at spatial infinity, with a suitable choice of co-ordinates, tend toward the values (4) ; i.e. we are considering gravitational fields which may be regarded as generated exclusively by matter in the finite region.

It might be thought that these approximations must lead us to Newton's theory. But to that end we still need to approximate the fundamental equations from a second point of view. We give our attention to the motion of a material point in accordance with the equations (16). In the case of the special theory of relativity the components

$$\frac{dx_1}{ds}, \frac{dx_2}{ds}, \frac{dx_3}{ds}$$

may take on any values. This signifies that any velocity

$$v = \sqrt{\left(\frac{dx_1}{dx_4}\right)^2 + \left(\frac{dx_2}{dx_4}\right)^2 + \left(\frac{dx_3}{dx_4}\right)^2}$$

may occur, which is less than the velocity of light *in vacuo*. If we restrict ourselves to the case which almost exclusively offers itself to our experience, of $v$ being small as compared with the velocity of light, this denotes that the components

$$\frac{dx_1}{ds}, \frac{dx_2}{ds}, \frac{dx_3}{ds}$$

are to be treated as small quantities, while $dx_4/ds$, to the second order of small quantities, is equal to one. (Second point of view of approximation.)

Now we remark that from the first point of view of approximation the magnitudes $\Gamma^{\tau}_{\mu\nu}$ are all small magnitudes of at least the first order. A glance at (46) thus shows that in this equation, from the second point of view of approximation, we have to consider only terms for which $\mu = v = 4$. Restricting ourselves to terms of lowest order we first obtain in place of (46) the equations

$$\frac{d^2 x_\tau}{dt^2} = \Gamma^\tau_{44}$$

where we have set $ds = dx_4 = dt$; or with restriction to terms which from the first point of view of approximation are of first order:—

$$\frac{d^2 x_\tau}{dt^2} = [44, \tau] \quad (\tau = 1, 2, 3)$$

$$\frac{d^2 x_4}{dt^2} = -[44, 4].$$

If in addition we suppose the gravitational field to be a quasi-static field, by confining ourselves to the case where the motion of the matter generating the gravitational field is but slow (in comparison with the velocity of the propagation of light), we may neglect on the right-hand side differentiations with respect to the time in comparison with those with respect to the space co-ordinates, so that we have

$$\frac{d^2 x_\tau}{dt^2} = -\tfrac{1}{2}\frac{\partial g_{44}}{\partial x_\tau} \quad (\tau = 1, 2, 3) \tag{67}$$

This is the equation of motion of the material point according to Newton's theory, in which $\tfrac{1}{2}g_{44}$ plays the part of the gravitational potential. What is remarkable in this result is that the component $g_{44}$ of the fundamental tensor alone defines, to a first approximation, the motion of the material point.

We now turn to the field equations (53). Here we have to take into consideration that the energy-tensor of "matter "is almost exclusively defined by the density of matter in the narrower sense, i.e. by the second term of the right-hand side of (58) [or, respectively, (58a) or (58b)]. If we form the approximation in question, all the components vanish with the one exception of $T_{44} = \rho = T$. On the left-hand side of (53) the second term is a small quantity of second order; the first yields, to the approximation in question,

$$\frac{\partial}{\partial x_1}[\mu v, 1] + \frac{\partial}{\partial x_2}[\mu v, 2] + \frac{\partial}{\partial x_3}[\mu v, 3] - \frac{\partial}{\partial x_4}[\mu v, 4].$$

For $\mu = v = 4$, this gives, with the omission of terms differentiated with respect to time,

$$-\tfrac{1}{2}\left(\frac{\partial^2 g_{44}}{\partial x_1^2} + \frac{\partial^2 g_{44}}{\partial x_2^2} + \frac{\partial^2 g_{44}}{\partial x_3^2}\right) = -\tfrac{1}{2}\nabla^2 g_{44}.$$

The last of equations (53) thus yields

$$\nabla^2 g_{44} = \kappa\rho \tag{68}$$

The equations (67) and (68) together are equivalent to Newton's law of gravitation.

By (67) and (68) the expression for the gravitational potential becomes

$$-\frac{\kappa}{8\pi}\int\frac{\rho d\tau}{r} \tag{68a}$$

while Newton's theory, with the unit of time which we have chosen, gives

$$-\frac{K}{c^2}\int\frac{\rho d\tau}{r}$$

in which K denotes the constant $6.7 \times 10^{-8}$, usually called the constant of gravitation. By comparison we obtain

$$\kappa = \frac{8\pi K}{c^2} = 1.87 \times 10^{-27} \tag{69}$$

## § 22. Behaviour of Rods and Clocks in the Static Gravitational Field. Bending of Light-rays. Motion of the Perihelion of a Planetary Orbit

To arrive at Newton's theory as a first approximation we had to calculate only one component, $g_{44}$, of the ten $g_{\mu\nu}$ of the gravitational field, since this component alone enters into the first approximation, (67), of the equation for the motion of the material point in the gravitational field. From this, however, it is already apparent that other components of the $g_{\mu\nu}$ must differ from the values given in (4) by small quantities of the first order. This is required by the condition $g = -1$.

For a field-producing point mass at the origin of co-ordinates, we obtain, to the first approximation, the radially symmetrical solution

$$\left.\begin{aligned} g_{\rho\sigma} &= -\delta_{\rho\sigma} - a\frac{x_\rho x_\sigma}{r^3} \quad (\rho, \sigma = 1, 2, 3) \\ g_{\rho4} &= g_{4\rho} = 0 \quad\quad (\rho = 1, 2, 3) \\ g_{44} &= 1 - \frac{a}{r} \end{aligned}\right\} \tag{70}$$

where $\delta_{\rho\sigma}$ is 1 or 0, respectively, accordingly as $\rho = \sigma$ or $\rho \neq \sigma$, and $r$ is the quantity $+\sqrt{x_1^2 + x_2^2 + x_3^2}$. On account of (68a)

$$a = \frac{\kappa M}{4\pi}, \tag{70a}$$

if M denotes the field-producing mass. It is easy to verify that the field equations (outside the mass) are satisfied to the first order of small quantities.

We now examine the influence exerted by the field of the mass M upon the metrical properties of space. The relation

$$ds^2 = g_{\mu\nu}dx_\mu dx_\nu.$$

always holds between the "locally" (§ 4) measured lengths and times $ds$ on the one hand, and the differences of co-ordinates $dx_\nu$ on the other hand.

For a unit-measure of length laid "parallel" to the axis of $x$, for example, we should have to set $ds^2 = -1$; $dx_2 = dx_3 = dx_4 = 0$. Therefore $-1 = g_{11}dx_1^2$. If, in addition, the unit-measure lies on the axis of $x$, the first of equations (70) gives

$$g_{11} = -\left(1 + \frac{a}{r}\right).$$

From these two relations it follows that, correct to a first order of small quantities,

$$dx = 1 - \frac{a}{2r} \tag{71}$$

The unit measuring-rod thus appears a little shortened in relation to the system of co-ordinates by the presence of the gravitational field, if the rod is laid along a radius.

In an analogous manner we obtain the length of co-ordinates in tangential direction if, for example, we set

$$ds^2 = -1; dx_1 = dx_3 = dx_4 = 0; x_1 = r, x_2 = x_3 = 0.$$

The result is

$$-1 = g_{22}dx_2^2 = -dx_2^2 \tag{71a}$$

With the tangential position, therefore, the gravitational field of the point of mass has no influence on the length of a rod.

Thus Euclidean geometry does not hold even to a first approximation in the gravitational field, if we wish to take one and the same rod, independently of its place and orientation, as a realization of the same interval; although, to be sure, a glance at (70a) and (69) shows that the deviations to be expected are much too slight to be noticeable in measurements of the earth's surface.

Further, let us examine the rate of a unit clock, which is arranged to be at rest in a static gravitational field. Here we have for a clock period $ds = 1$; $dx_1 = dx_2 = dx_3 = 0$ Therefore

$$1 = g_{44}dx_4^2;$$

$$dx_4 = \frac{1}{\sqrt{g_{44}}} = \frac{1}{\sqrt{(1+(g_{44}-1))}} = 1 - \tfrac{1}{2}(g_{44} - 1)$$

or

$$dx_4 = 1 + \frac{\kappa}{8\pi}\int \rho\frac{d\tau}{r} \tag{72}$$

Thus the clock goes more slowly if set up in the neighborhood of ponderable masses. From this it follows that the spectral lines of light reaching us from the surface of large stars must appear displaced towards the red end of the spectrum.[14]

We now examine the course of light-rays in the static gravitational field. By the special theory of relativity the velocity of light is given by the equation

$$- dx_1^2 - dx_2 - dx_3^2 + dx_4^2 = 0$$

and therefore by the general theory of relativity by the equation

$$ds^2 = g_{\mu\nu} dx_\mu dx_\nu = 0 \tag{73}$$

If the direction, i.e. the ratio $dx_1 : dx_2 : dx_3$ is given, equation (73) gives the quantities

$$\frac{dx_1}{dx_4}, \frac{dx_2}{dx_4}, \frac{dx_3}{dx_4}$$

and accordingly the velocity

$$\sqrt{\left(\frac{dx_1}{dx_4}\right)^2 + \left(\frac{dx_2}{dx_4}\right)^2 + \left(\frac{dx_3}{dx_4}\right)^2} = \gamma$$

defined in the sense of Euclidean geometry. We easily recognize that the course of the light-rays must be bent with regard to the system of co-ordinates, if the $g_{\mu\nu}$ are not constant. If $n$ is a direction perpendicular to the propagation of light, the Huyghens principle shows that the light-ray, envisaged in the plane ($\gamma$, $n$), has the curvature $-\partial\gamma/\partial n$.

We examine the curvature undergone by a ray of light passing by a mass M at the distance $\Delta$. If we choose the system of co-ordinates in agreement with the accompanying diagram, the total bending of the ray (calculated positively if concave towards the origin) is given in sufficient approximation by

$$\text{B} = \int_{-\infty}^{+\infty} \frac{\partial\gamma}{\partial x_1} dx_2,$$

while (73) and (70) give

$$\gamma = \left(-\frac{g_{44}}{g_{22}}\right) = 1 - \frac{a}{2r}\left(1 + \frac{x_2^2}{r^2}\right).$$

Carrying out the calculation, this gives

---

14 According to E. Freundlich, spectroscopical observations on fixed stars of certain types indicate the existence of an effect of this kind, but a crucial test of this consequence has not yet been made.

$$B = \frac{2a}{\Delta} = \frac{\kappa M}{2\pi\Delta} \tag{74}$$

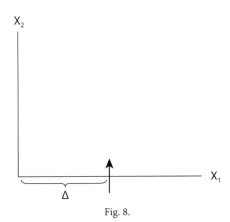

Fig. 8.

According to this, a ray of light going past the sun undergoes a deflection of 1.7″; and a ray going past the planet Jupiter a deflection of about .02″.

If we calculate the gravitational field to a higher degree of approximation, and likewise with corresponding accuracy the orbital motion of a material point of relatively infinitely small mass, we find a deviation of the following kind from the Kepler-Newton laws of planetary motion. The orbital ellipse of a planet undergoes a slow rotation, in the direction of motion, of amount

$$\varepsilon = 24\pi^3 \frac{a^2}{T^2 c^2 (1-e^2)} \tag{75}$$

per revolution. In this formula $a$ denotes the major semi-axis, $c$ the velocity of light in the usual measurement, $e$ the eccentricity, $T$ the time of revolution in seconds.[15]

Calculation gives for the planet Mercury a rotation of the orbit of 43″ per century, corresponding exactly to astronomical observation (Leverrier); for the astronomers have discovered in the motion of the perihelion of this planet, after allowing for disturbances by other planets, an inexplicable remainder of this magnitude.

---

15  For the calculation I refer to the original papers: A. Einstein, *Sitzungsber. d. Preuss. Akad. d. Wiss.*, 1915, p. 831; K. Schwarzschild, ibid., 1916, p. 189.

# HAMILTON'S PRINCIPLE AND THE GENERAL THEORY OF RELATIVITY

by A. Einstein

H. A. LORENTZ AND D. HILBERT HAVE RECENTLY SUCCEEDED[1] IN PRESENTING THE theory of general relativity in a particularly comprehensive form by deriving its equations from a single variational principle. The same shall be done in this paper. My aim here is to present the fundamental connections as transparently and comprehensively as the principle of general relativity allows. In contrast to HILBERT's presentation, I shall make as few assumptions about the constitution of matter as possible. On the other hand, and in contrast to my own very recent treatment of the subject matter, the choice of a system of coordinates shall remain completely free.

## §1. THE VARIATIONAL PRINCIPLE AND THE FIELD EQUATIONS OF GRAVITATION AND MATTER

The gravitational field shall be described as usual by the tensor[2] of the $g_{\mu\nu}$ (or $g^{\mu\nu}$ resp.); matter (inclusive of the electromagnetic field) by an arbitrary number of space-time functions $q_{(\rho)}$ whose invariance-theoretical character we ignore. Furthermore, let $\mathfrak{H}$ be a function of the[{1}]

$$g^{\mu\nu}, g^{\mu\nu}_\sigma \left( = \frac{\partial g^{\mu\nu}}{\partial x_\sigma} \right) \text{ and } g^{\mu\nu}_{\sigma\tau} \left( = \frac{\partial^2 g^{\mu\nu}}{\partial x_\sigma \partial x_\tau} \right), \text{ the } q_{(\rho)} \text{ and } q_{(\rho)\alpha} \left( = \frac{\partial q_{(\rho)}}{\partial x_\alpha} \right).$$

---

[1]    Four papers by H. A. LORENTZ in volumes 1915 and 1916 of the *Publikationer d. Koninkl. Akad van Wetensch. te Amsterdam*; D. HILBERT, *Gött. Nachr.* 1915, Heft. 3.

[2]    For the time being, the tensorial character of the $g_{\mu\nu}$ is not used.

The variational principle

$$\delta\left\{\int \mathfrak{H} d\tau\right\} = 0 \tag{1}$$

then provides as many differential equations as there are functions $g_{\mu\nu}$ and $q_{(\rho)}$, which are to be determined, provided we agree that the $g^{\mu\nu}$ and $q_{(\rho)}$ are to be varied independently of each other such that at the boundaries of integration the $\delta q_{(\rho)}$, $\delta g^{\mu\nu}$, and $\partial \delta g_{\mu\nu}/\partial x_\sigma$ all vanish.

We shall now assume $\mathfrak{H}$ to be linear in the $g_{\sigma\tau}^{\mu\nu\{2\}}$ such that the coefficients of the $g_{\sigma\tau}^{\mu\nu}$ depend only upon the $g^{\mu\nu}$. The variational principle (1) can then be replaced by one more convenient for us. With suitable partial integration one gets

$$\int \mathfrak{H} d\tau = \int \mathfrak{H}^* d\tau + F, \tag{2}$$

where $F$ is an integral extended over the boundaries of the domain under consideration, while the quantity $\mathfrak{H}^*$ depends only upon the $g^{\mu\nu}$, $g_\sigma^{\mu\nu}$, $q_{(\rho)}$, $q_{(\rho)\alpha}$ but no longer upon $g_{\sigma\tau}^{\mu\nu}$. For the variation of interest to us one gets from (2)

$$\delta\left\{\int \mathfrak{H} d\tau\right\} = \delta\left\{\int \mathfrak{H}^* d\tau\right\}, \tag{3}$$

whereupon we can replace the variational principle (1) with the more convenient one

$$\delta\left\{\int \mathfrak{H}^* d\tau\right\} = 0. \tag{1a}$$

By executing the variation after the $g^{\mu\nu}$ and the $q_{(\rho)}$ one obtains for the field equations of gravitation and matter the equations[3]

$$\frac{\partial}{\partial x_\alpha}\left(\frac{\partial \mathfrak{H}^*}{\partial g_\alpha^{\mu\nu}}\right) - \frac{\partial \mathfrak{H}^*}{\partial g^{\mu\nu}} = 0 \tag{4}$$

$$\frac{\partial}{\partial x_\alpha}\left(\frac{\partial \mathfrak{H}^*}{\partial q_{(\rho)\alpha}}\right) - \frac{\partial \mathfrak{H}^*}{\partial q_{(\rho)}} = 0. \tag{5}$$

## §2. SEPARATE EXISTENCE OF THE GRAVITATIONAL FIELD

The energy components cannot be split into two separate parts such that one belongs to the gravitational field and the other to matter, unless one makes special assumptions in which manner $\mathfrak{H}$ should depend upon the $g^{\mu\nu}$, $g_{\sigma\tau}^{\mu\nu}$, $q_{(\rho)}$, $q_{(\rho)\alpha}$. In order to bring about this property of the theory we assume

$$\mathfrak{H} = \mathfrak{G} + \mathfrak{M}, \tag{6}$$

---

[3]    As an abbreviation, the summation signs are omitted in the formulas. A summation has to be carried out over the indices that occur twice in a term. For example, in (4) $\partial/\partial x_\alpha(\partial \mathfrak{H}^*/\partial g_\alpha^{\mu\nu})$ denotes the term $\sum_\alpha \partial/\partial x_\alpha(\partial \mathfrak{H}^*/\partial g_\alpha^{\mu\nu})$.

Where $\mathfrak{G}$ depends only upon $g^{\mu\nu}$, $g_\sigma^{\mu\nu}$, $g_{\sigma\tau}^{\mu\nu}$, and $\mathfrak{M}$ only upon $g^{\mu\nu}$, $q_{(\rho)}$, $q_{(\rho)\alpha}$.

Equations (4), (5)[3] then take the form

$$\frac{\partial}{\partial x_\alpha}\left(\frac{\partial\mathfrak{G}^*}{\partial g_\alpha^{\mu\nu}}\right) - \frac{\partial\mathfrak{G}^*}{\partial g^{\mu\nu}} = \frac{\partial\mathfrak{M}}{\partial g^{\mu\nu}} \tag{7}$$

$$\frac{\partial}{\partial x_\alpha}\left(\frac{\partial\mathfrak{M}}{\partial q_{(\rho)\alpha}}\right) - \frac{\partial\mathfrak{M}}{\partial q_{(\rho)}} = 0 \tag{8}$$

$\mathfrak{G}^*$ is here in the same relation to $\mathfrak{G}$ as $\mathfrak{H}^*$ is to $\mathfrak{H}$.

It must be noted that equations (8) or (5) respectively would have to be replaced by others if we would assume that $\mathfrak{M}$ or $\mathfrak{H}$ resp. would depend upon higher than the first derivatives of $q_{(\rho)}$. Similarly, one could imagine the $q_{(\rho)}$ not as mutually independent but rather as connected to each other by further, conditional equations. All this is irrelevant for the following development, since it is solely based upon equation (7), which is obtained by varying our integral after the $g^{\mu\nu}$.[4]

## §3. PROPERTIES OF THE FIELD EQUATIONS OF GRAVITATION BASED ON THE THEORY OF INVARIANTS

We now introduce the assumption that

$$ds^2 = g_{\mu\nu}dx_\mu dx_\nu \tag{9}$$

is an invariant. This fixes the transformational character of the $g_{\mu\nu}$. We make no presuppositions about the transformational character of the $q_{(\rho)}$ which describe matter. However, the functions $H = \mathfrak{H}/\sqrt{-g}$ and $G = \mathfrak{G}/\sqrt{-g}$ and $M = \mathfrak{M}/\sqrt{-g}$ shall be invariants under arbitrary substitutions of the space-time coordinates. From these suppositions follows the general covariance of equations (7) and (8), which have been derived from (1). It follows furthermore that (up to a constant factor) $G$ is equal to the scalar of the RIEMANN tensor of curvature, because there is no other invariant with the properties demanded for $G$.[4] With this, $\mathfrak{G}^*$, and hence the left-hand side of the field equation (7) is completely determined.[5]

The postulate of general relativity entails certain properties of the function $\mathfrak{G}^*$ which we shall now derive. For this purpose we carry out an infinitesimal transformation of the coordinates by setting

$$x_\nu' = x_\nu + \Delta x_\nu, \tag{10}$$

---

4    This is the reason why the requirements of general relativity led to a quite distinct theory of gravitation.

5    Execution of the partial integration yields

$$\mathfrak{G}^* = \sqrt{-g}\,g^{\mu\nu}\left[\begin{Bmatrix}\mu\alpha\\\beta\end{Bmatrix}\begin{Bmatrix}\mu\beta\\\alpha\end{Bmatrix} - \begin{Bmatrix}\mu\nu\\\alpha\end{Bmatrix}\begin{Bmatrix}\alpha\beta\\\beta\end{Bmatrix}\right].$$

where the $\Delta x_v$ are arbitrarily eligible, infinitesimally small functions of the coordinates. $x'_v$ are the coordinates of the world point in the new system, the same point whose coordinates were $x_v$ in the original system. Just as for the coordinates, there is a transformation law for any other quantity $\psi$, of the type

$$\psi' = \psi + \Delta\psi,$$

where $\Delta\psi$ must always be expressible in terms of the $\Delta x_v$. From the covariant properties of the $g^{\mu\nu}$ one derives easily for the $g^{\mu\nu}$ and the $g^{\mu\nu}_\sigma$ the transformation laws:

$$\Delta g^{\mu\nu} = g^{\mu\alpha}\frac{\partial \Delta x_\nu}{\partial x_\alpha} + g^{\nu\alpha}\frac{\partial \Delta x_\mu}{\partial x_\alpha} \tag{11}$$

$$\Delta g^{\mu\nu}_\sigma = \frac{\partial(\Delta g^{\mu\nu})}{\partial x_\sigma} - g^{\mu\nu}_\alpha\frac{\partial \Delta x_\alpha}{\partial x_\sigma}. \tag{12}$$

$\Delta\mathfrak{G}^*$ can be calculated with the help of (11) and (12)[5], since $\mathfrak{G}^*$ depends only upon the $g^{\mu\nu}$ and the $g^{\mu\nu}_\sigma$. Thus, one gets the equation

$$\sqrt{-g}\,\Delta\left(\frac{\mathfrak{G}^*}{\sqrt{-g}}\right) = S^\nu_\sigma\frac{\partial \Delta x_\sigma}{\partial x_\nu} + 2\frac{\partial \mathfrak{G}^*}{\partial g^{\mu\nu}_\alpha}g^{\mu\nu}\frac{\partial^2 \Delta x_\sigma}{\partial x_\nu \partial x_\alpha}, \tag{13}$$

where we used the abbreviation

$$S^\nu_\sigma = 2\frac{\partial \mathfrak{G}^*}{\partial g^{\mu\sigma}}g^{\mu\nu} + 2\frac{\partial \mathfrak{G}^*}{\partial g^{\mu\sigma}_\alpha}g^{\mu\nu}_\alpha + \mathfrak{G}^*\delta^\nu_\sigma - \frac{\partial \mathfrak{G}^*}{\partial g^{\mu\alpha}_\nu}g^{\mu\alpha}_\sigma. \tag{14}$$

From these two equations we draw two conclusions that are important in the following. We know $\mathfrak{G}/\sqrt{-g}$ to be an invariant under arbitrary substitutions but not $\mathfrak{G}^*/\sqrt{-g}$. It is, however, easy to show that the latter quantity is invariant under *linear* substitutions of the coordinates. Consequently, the right-hand side of (13) must always vanish when all $\partial^2\Delta x_\sigma/\partial x_\nu\partial x_\alpha$ do. Then it follows that $\mathfrak{G}^*$ must satisfy the identity

$$S^\nu_\sigma \equiv 0. \tag{15}$$

If we furthermore choose the $\Delta x_v$ such that they differ from zero only inside the domain considered, but vanish in an infinitesimal neighborhood of the boundary, then the value of the integral in equation (2) extended over the boundary does not change during the transformation. We therefore have

$$\Delta(F) = 0$$

and thus[6]

---

[6] Introducing $\mathfrak{G}$ and $\mathfrak{G}^*$ instead of $\mathfrak{H}$ and $\mathfrak{H}^*$.

$$\Delta\left\{\int \mathfrak{G}d\tau\right\} = \Delta\left\{\int \mathfrak{G}^{*}d\tau\right\}.$$

But the left-hand side of the equation must vanish since both $\mathfrak{G}/\sqrt{-g}$ and $\sqrt{-g}\,d\tau$ are invariants. Consequently, the right-hand side vanishes also. Due to (13), (14), and (15)[6] we next get the equation

$$\int \frac{\partial \mathfrak{G}^{*}}{\partial g_{\alpha}^{\mu\sigma}} g^{\mu\nu} \frac{\partial^{2}\Delta x_{\sigma}}{\partial x_{\nu}\partial x_{\alpha}}\,d\tau = 0. \tag{16}$$

Rearranging after twofold partial integration, and considering the free choice of the $\Delta x_{\sigma}$, one has the identity

$$\frac{\partial^{2}}{\partial x_{\nu}\partial x_{\alpha}}\left(\frac{\partial \mathfrak{G}^{*}}{\partial g_{\alpha}^{\mu\sigma}} g^{\mu\nu}\right) \equiv 0. \tag{17}$$

We now have to draw conclusions from the two identities (15)[7] and (17), which follow from the invariance of $\mathfrak{G}/\sqrt{-g}$, i.e., from the postulate of general relativity.

The field equations (7) of gravitation can be transformed first by mixed multiplication with $g^{\mu\nu}$. One obtains then (also exchanging the indices $\sigma$ and $\nu$) as an equivalent of the field equations (7) the equations

$$\frac{\partial}{\partial x_{\alpha}}\left(\frac{\partial \mathfrak{G}^{*}}{\partial g_{\alpha}^{\mu\sigma}} g^{\mu\nu}\right) = -(\mathfrak{T}_{\sigma}^{\nu} + \mathbf{t}_{\sigma}^{\nu}), \tag{18}$$

where we put

$$\mathfrak{T}_{\sigma}^{\nu} = -\frac{\partial \mathfrak{M}}{\partial g^{\mu\sigma}} g^{\mu\nu} \tag{19}$$

$$\mathbf{t}_{\sigma}^{\nu} = -\left(\frac{\partial \mathfrak{G}^{*}}{\partial g_{\alpha}^{\mu\sigma}} g_{\alpha}^{\mu\nu} + \frac{\partial \mathfrak{G}^{*}}{\partial g^{\mu\sigma}} g^{\mu\nu}\right) = \frac{1}{2}\left(\mathfrak{G}^{*}\delta_{\sigma}^{\nu} - \frac{\partial \mathfrak{G}^{*}}{\partial g_{\nu}^{\mu\alpha}} g_{\sigma}^{\mu\alpha}\right). \tag{20}$$

The latter expression for $\mathbf{t}_{\sigma}^{\nu}$ is justified by (14) and (15). After differentiation of (18) with respect to $x_{n}$ and summation over $\nu$, with consideration of (17), follows

$$\frac{\partial}{\partial x_{\nu}}(\mathfrak{T}_{\sigma}^{\nu} + \mathbf{t}_{\sigma}^{\nu}) = 0. \tag{21}$$

Equation (21) expresses the conservation of the momentum and the energy. We call $\mathfrak{T}_{\sigma}^{\nu}$ the components of the energy of matter, $\mathbf{t}_{\sigma}^{\nu}$ the components of the energy of the gravitational field.

From the field equations (7) of gravitation follows (after multiplication by $g_{\sigma}^{\mu\nu}$, summation over $\mu$ and $\nu$, and on account of (20))

$$\frac{\partial \mathbf{t}_{\sigma}^{\nu}}{\partial x_{\nu}} + \frac{1}{2} g_{\sigma}^{\mu\nu} \frac{\partial \mathfrak{M}}{\partial g^{\mu\nu}} = 0,$$

or, taking (19) and (21) into account,

$$\frac{\partial \mathfrak{T}_\sigma^v}{\partial x_v} + \frac{1}{2} g_\sigma^{\mu v} \mathfrak{T}_{\mu v} = 0,\tag{22}{\{8\}}$$

where $\mathfrak{T}_{\mu v}$ denotes the quantities $g_{v\sigma}\mathfrak{T}_\mu^\sigma$. These are four equations that the energy components of matter have to satisfy.

It is to be emphasized that the (generally covariant) conservation theorems (21) and (22) have been derived—using also the postulate of general covariance (relativity)—from the field equations (7) of gravitation *alone*, without use of the field equations (8) for material processes.

### Additional notes by translator

In his footnote 1), just prior to equations (4) and (5), Einstein introduces into tensor calculus, in a formal manner, the rule of abbreviated writing of summations, which is now generally known as the Einstein summation convention. It was introduced in Doc. 30, p. 296.

{1}   The "$q$" with 2 subscripts and 2 superscripts has been corrected here to "$g$"; the 2nd derivative of "$q$" inside the following parenthesis has also been corrected to "$g$," both with indices as indicated. Editorial notes [6] and [7] relate to similar typesetting errors.

{2}   "$q_{\sigma\tau}^{\mu v}$" has been corrected here to "$g_{\sigma\tau}^{\mu v}$."

{3}   "(4a)" has been corrected to "(5)."

{4}   "$q^{\mu v}$" has been corrected here to "$g^{\mu v}$".

{5}   "(13)" and "(14)" have been corrected here to "(11)" and "(12)."

{6}   "(14), (15), (16)" have been corrected here to "(13), (14), (15)."

{7}   "(16)" has been corrected to "(15)."

{8}   "–"preceding the factor $\frac{1}{2}$ has been corrected here to "+".

# INDEX

Abraham, Max (Abraham's theory of gravitation), 9, 14–15, 160, 165
acceleration, 14–16, 23, 45, 47, 53, 75, 93, 167, 185, 186, 188. *See also* frame (system) of reference; accelerated
affine connection, 77, 81, 89
Albert Einstein Archives (Hebrew University), 2, 5, 179
Ampère's law, 111
astronomical tests of general relativity, 26, 117, 123, 125, 127, 152, 160
atoms: absorption of radiation by, 139; emission of radiation by, 139; the existence of, 7
*Autobiographical Notes* (Einstein), 43, 101, 113

Benjamin, Walter, 1
Berliner, Arnold, 3
Bernays, Paul, 26, 165
Besso, Michele, 9, 11, 25, 30, 34, 59, 121, 129, 135, 137, 139, 161, 165
Born, Max, 17, 166
boundary conditions, 151

Cartan, Elie, 23, 163
Charles University (Prague), 8, 53
Christoffel, Elwin Bruno, 19, 32, 39, 67, 77, 166
Christoffel symbol 29, 77, 87, 89, 93, 95; geometric and physical significance of, 75
clocks: "atomic clocks," 127; behavior of rods and clocks in the static gravitation field, 223–226; Einstein on the behavior of clocks in a gravitational field, 49, 125, 127
connection, 77. *See also* affine connection
conservation, relation of to covariance, 107
conservation principle. *See* energy-momentum conservation

coordinates: coordinate condition, 31, 71; coordinate restriction, 31, 71; unimodular coordinates, 33, 34, 71, 95
correspondence principle, 21, 22, 24, 27, 63; coordinate conditions that follow from it, 31
"Cosmological Considerations in the General Theory of Relativity" (Einstein), 151, 162
cosmological constant, 151–152, 154; Einstein's "biggest blunder" 152 (box), 162, 163
cosmology, development of modern, 151–154
covariance. *See* general covariance
covariant differentiation, 75, 77, 79, 81, 87, 89
curvature, 89
curved surfaces, Gaussian theory of, 19, 39, 57, 59, 73

de Sitter, Willem, 135, 152, 163, 166, 176
differential calculus, 22, 25, 32, 57, 183. *See also* tensor calculus
divergence, 216, 221; of a contravariant vector, 204–205, 206; mathematical concept of, 85; of a mixed tensor of the second rank, 87, 207–208; vanishing of divergence of the material energy-tensor, 217; of a vector field, 85
dynamics, 97; relativistic continuum dynamics, 115

Eddington, Arthur, 150, 155, 166
Ehrenfest, Paul, 17, 33, 97, 123, 151, 162, 167
"Ehrenfest paradox," 18
Einstein, Albert, 1–5, 7, 10–11, 26, 33–34, 83, 97, 107–109, 115–119, 125–131, 135–139, 149; academic career of, 8–12; collaboration with Besso (*see* Besso, Michele); collaboration with Grossmann

Einstein, Albert (*continued*)
(*see* Grossmann, Marcel); fascination with Maxwell theory, heuristics, 15–17, 20–26, 35, 65, 69, 83, 95, 103, 149–151, 153, 154; on the nature of matter, 29–30; Princeton lectures, 81; relations with Hilbert (*see* Hilbert, David)
Einstein, Elsa, 4, 12
Einstein, Hans Albert, 105
Einstein Papers Project (California Institute of Technology), 5
Einstein tensor, 31, 103
Einstein Tower, 2
Einstein-Besso manuscript, the, 10, 11, 26
Eisenstaedt, Jean, 155
"The Electrodynamics of Moving Bodies, The" (Einstein [1905]), history of the manuscript of, 1–5, 159
electromagnetism (electrodynamics, electromagnetic field), 9, 13, 20, 21, 47, 61, 67, 85, 109, 217, 219, 220; electrodynamic theory of matter, 99, 109; energy-momentum components of, 115, 220–222; Lorentz model of, 21, 61; Maxwell's electromagnetic field equations, 111, 113, 217–221
electromagnetic radiation and matter, interaction between, 139
energy-momentum: of the gravitational field, 99, 101, 211–215; laws of for matter as a consequence of the field equations, 216–217; of matter, 101. *See also* energy-momentum conservation principle/law
energy-momentum conservation, 21–23, 26, 63, 65, 71, 83, 85, 87, 97, 99, 105, 107, 111, 115, 131, 135, 139, 214–218, 231; in the absence of matter, 99; coordinate restrictions that follow from, 31; Einstein's proof of for the electromagnetic field, 111; laws of in the general case, 215–216; mathematical formulation of in general relativity, 87; symmetries in nature and, 107
energy-momentum tensor, 63, 87, 103, 109, 115, 160, 210–215
*Entwurf* paper/theory, the ("Outline of a Generalized Theory of Relativity and of a Theory of Gravitation" [Einstein and Grossman]), 10, 11, 12, 24–26, 29, 30, 33, 51, 71, 91, 93, 97, 99, 107, 161; abandonment of the theory by Einstein 12, 25, 27–28, an intermediate step toward the general theory of relativity, 83; mathematical formalism of, 29; use of in explaining the precession of the perihelion of Mercury, 129
Eötvös, Lorand, 45, 167

equivalence principle, the, 8, 15–17, 21, 22, 23, 45, 69, 159, 160; phenomena derived from, 117, 127; and the rate of clocks, 49; and the relativistic theory of gravitation, 53, 63; and special relativity, 95
ether, 39, 113, 155
"Ether and the Theory of Relativity" (Einstein), 113
Euler, Leonhard, 109, 167; equations in hydrodynamics of, 217–218
"Explanation of Perihelion Motion of Mercury from the General Theory of Relativity" (Einstein), 30–31

Faraday, Michael, 101, 111
Faraday's law, 111, 113
field equation. *See* gravitational field equation
field theory. *See* electromagnetism; gravitational field equation
"Formal Foundation of the General Theory of Relativity, The" (Einstein [1914]), 12, 33, 109, 111, 161
"Foundation of General Relativity" (appendix of ["Formulation of the Theory on the Basis of a Variational Principle"]), 33–34, 131, 162; Einstein's original decision not to publish the appendix, 131; publication of a modified version of the appendix, 133
"Foundation of General Relativity" (text of [Einstein, 1916]), 2, 77, 149, 162, 183, 185–186
frame (system) of reference: inertial, accelerated, 15–19, 23, 39, 45, 53, 63, 67, 83, 113, 135, 187; rotating, 15, 17–18, 26–27, 43, 47, 49, 59, 81, 121, 160. *See also* acceleration; rotation
Freundlich, Erwin, 2–3, 4, 123, 127, 160, 168
Friedmann, Alexander, 153, 163, 168
Fokker, Adrian D., 161

Galilei, Galileo, 14, 185
Galileo's principle, 14–16, 45
Gamow, George, 152 (box), 168
Gauss, Carl Friedrich, 32, 39, 47, 169
Gauss's law, 111, 113
general covariance, 11, 24, 25, 28, 49, 51, 57, 65, 69, 83, 93, 105, 121, 135, 187–189, 232; relation of to conservation, 107
geodetic (geodesic), 19, 55, 73, 79, 81, 93, 209; derivation of the geodetic line equation by means of the "variational method," 73, 75; equation of the geodetic line and the motion of a particle, 198–201; relation of to the concept of

"affine connection," 77; as the straightest possible line, 77

geometry: differential geometry, 67, 75, 81; Einstein's view of geometry as an ancient branch of physics, 49; Euclidian geometry, 18, 19, 47, 187, 188; Gaussian surface geometry, 39, 57, 59, 73; non-Euclidian geometry, 8, 47, 81; practical geometry, 49; Riemannian geometry, 10, 57, 65, 73, 75, 77; role of in physics, 17–20, 49; spacetime geometry, 53, 93. *See also* differential calculus

gravitation (gravity), 12–15, 18–20, 45, 93, 101, 234; effect on the rate of clocks, 127; field, 93; gravitational field and acceleration, 45; gravitational potential, 8–9, 22, 55, 91, 117, 121; Newton's laws/theory of, 7, 8, 21, 69, 95, 105; relativity of, 15–16. *See also* Abraham's theory of gravitation; gravitational field equation; Nordstrom's theory of gravitation

gravitational field equation, 21, 22–23, 24, 31, 65, 93, 99, 103, 131, 139, 151, 161, 162, 183, 234; in the absence of matter, 209–214; inclusion of matter in, 101; the "November field equation," 105; and the variational principle, 229–230; theory of the gravitation field, 210–211. *See also* "Foundation of General Relativity" (Appendix of ["Formulation of the Theory on the Basis of a Variational Principle"]); "Hamilton's Principle and the General Theory of Relativity" (Einstein)

gravitational redshift, 33, 117, 123, 127, 160

gravitational waves, 135,156, 162

Grossman, Marcel, 10, 19, 22, 24, 25, 26, 28, 39, 57, 59, 67, 69, 83, 105, 161, 169, 183

Habicht, Conrad, 26, 43, 129, 159

Hale, George, 123, 127, 169

Hamilton, William Rowan, 97. *See also* variational principle (formalism)

Hamiltonian function (formalism), 55, 97, 131, 133, 135, 137, 211–214. *See also* Lagrangian function (formalism)

"The Hamiltonian Principle and the General Theory of Relativity," 33, 119, 131, 133, 135, 139, 162, 229–234

Hebrew (Jewish) University, 2–5, 179

Hertz, Heinrich, 20, 41, 101, 170

Hilbert, David, 31, 34, 107, 131, 133, 170, 227; axiomatic formulation of physics by, 99, 133; competition with Einstein, 32–33, 99, 107, 135

"hole argument," 25, 51, 161

Holton, Gerald, 12

"How I Created the Theory of Relativity" (Einstein), 45

Hubble, Edwin, 152, 153–154, 163, 170

Hume, David, 43, 170

hydrodynamics, 115, 216–217; and Einstein's theory of gravity, 109

inertial forces, 16–18, 63, 149

inertial (frame of reference) system. *See* frame (system) of reference

Janssen, Michel, 28

Jupiter, gravitational field of and light refraction, 123

Kepler, Johannes, 26, 129

Kohn, Leo, 4

Kollek, Theodore, 1

Kottler, Friedrich, 171

Kretschmann, Erich Justus, 162

Lagrangian function (formalism), 26, 97, 99, 131. *See also* Hamiltonian function (formalism)

Lampa, Anton, 8

Le Verrier, Urbain, 95, 129

Lemaître, Georges Henri, 153, 163, 171

Levi-Civita, Tullio, 10, 19, 22, 23, 32, 39, 67, 81, 89, 172

Levi-Civita connection, 77. *See also* affine connection

light: bending (deflection), 8, 9, 17, 33, 45, 115, 117, 123, 127, 150, 160, 162, 223; as a wave, 7, 39. *See also* light, velocity of

light, velocity of, 9, 13, 17, 39, 41, 113, 121, 187, 221, 225

Loewe, Heinrich, 3

Lorentz, Hendrik Antoon, 1, 10, 12, 20, 25, 34, 41, 83, 97, 111, 133, 135, 172, 227

Lorentz contraction, 18, 47, 187

Mach, Ernst, 8, 25, 26, 43, 160, 161, 172; critique of classical mechanics by, 15, 41, 150, 151—154

Mach's principle, 41, 125, 152, 153, 154

magnetism: magnetic fields, 20, 111; magnetic monopoles, 111. *See also* electromagnetism

Marić, Mileva, 8, 12

mass, 125, 127, 155, 186, 187, 214; as a form of energy, 7; gravitational mass, 14, 17, 45; inertial mass, 14, 15, 45, 115; mass density, 101, 103, 109, 115, 121

matrix, determinant of, 69

matter, 21, 29–30, 101, 103, 152; electro-dynamic theory of, 99

Maxwell, James Clerk, 13, 20, 39, 41, 101, 111, 173

Maxwell equations, 39, 61, 113, 217–220; effect of gravitation on, 111. *See also* electromagnetism

mechanics, 39, 152; classical (Newtonian) mechanics, 15, 41, 45, 51, 63, 73, 150, 153, 184–186

Mercury, precession of the perihelion orbit of, 26, 27, 28, 30–31, 95, 117, 159, 162; Einstein's explanation for, 129, 224–227; and general covariance, 121; Schwarzschild solution of, 121, 154

metric (fundamental), 22, 53, 55, 67, 69, 83,121, 205–206; the contravariant fundamental tensor, 197, 204; the covariant fundamental tensor, formation of new tensors by means of the fundamental tensor, 195, 199

Mie, Gustav, 14–15, 99, 109, 131, 173

Minkowski, Hermann, 8, 13, 61, 173, 183, 190; mathematical formalism of, 39, 53, 159, 160; metric of, 71

momentum. *See* energy-momentum conservation

Nernst, Walther, 11, 12, 174

"New Determination of Molecular Dimensions, A" (Einstein), 59

"New Formal Interpretation of Maxwell's Field Equations of Electrodynamics" (Einstein), 111

Newton, Isaac, 7, 14, 15, 185; and the laws of planetary motion, 26; Newtonian mechanics (physics), 19–20, 23, 41, 51, 75, 186; rotating bucket experiment of, 26, 41, 47. *See also* mechanics; gravitation (gravity), Newton's laws/theory of; Newtonian limit of general relativity

Newtonian limit of general relativity 23, 31, 107, 117, 121, 222–224

Noether, Emmy, 33–34, 107, 174

Noether's theorem, 33–34, 107

Nordström, Gunnar (Nordstrom's theory of gravitation), 125, 161, 174

"November theory," 29, 31, 107

"Olympia Academy," 43

"On the Generalized Theory of Relativity" (Einstein), 28–29, 161; addendum to, 29–30, 162

"On the Present State of the Problem of Gravitation" (Einstein), 125

Oppenheim, Paul, 3

"Outline of a Generalized Theory of Relativity and of a Theory of Gravitation." *See*

Entwurf paper/theory, the ("Outline of a Generalized Theory of Relativity and of a Theory of Gravitation" [Einstein and Grossman])

Parmitano, Luca, 4

physics: classical (Newtonian) physics, 7, 14, 17, 19–21, 23–24, 39, 41, 49, 53, 55, 63, 75, 85, 93, 111, 115; nature of the laws of, 43, 185, 186–187; role of geometry in, 17–20

Pirani, Felix, 154

Planck, Max, 11, 12

Poincaré, Henri, 159

Poisson equation, the, 69, 101, 103, 107

"Princeton lectures," 81, 163

Pythagorean theorem, extension of to four dimensions, 53

quantum theory, 139, 155

relativity, special theory of, 1, 7, 13—15, 17, 21, 39, 41, 45, 53, 61, 63, 95, 105, 113, 117, 137, 150, 159, 160, 183, 184, 186–190, 197, 208—209, 213

*"Relativity: The Special and General Theories (A Popular Account* [Einstein]*),"* or, "About the Special and General Theory of Relativity in Plain Terms (Einstein)," 9–10, 55, 59, 137, 154

Ricci-Curbastro, Gregorio, 10, 22, 32, 39, 67, 77, 81, 175

Ricci tensor, 26, 29, 31, 91, 95, 103

Riemann, Bernhard, 10, 19, 32, 39, 67, 175

Riemann (curvature, Riemann-Christoffel) tensor, 24, 26, 29, 31, 81, 89, 91, 95, 133, 162, 207–209

Riemannian geometry, 10, 57, 63, 65, 73, 77, 161

rods, behavior of in a gravitational field, 125, 223–226

rotating disk, 17, 18, 26, 47, 49, 81

rotation, 22–23, 26, 41,43, 47, 59, 85, 101, 129, 186. *See also* frame (system) of reference, rotating

Royal Prussian Academy of Science, 1, 11, 12, 33, 77, 95, 121, 131, 161; Einstein's November 1915 papers, 28–32, 97, 105

Safra, Edmond, 1

Schlick, Moritz, 51, 115, 175

Schwarzschild, Karl, 121, 123, 129, 154, 176

Schwarzschild solution, the, 121, 154–155, 162

six-vector, 61, 109, 193; antisymmetric extension of, 85, 205–206; divergence of, 85, 205–206

Solovine, Maurice, 43
Sommerfeld, Arnold, 28, 29, 33, 57, 176
spacetime, 9, 13, 15, 19, 20, 39, 51, 53, 55,
     67, 73, 113, 123, 133; curved spacetime,
     24, 49, 67, 69; Minkowski spacetime, 53,
     69, 71, 89; spacetime geometry, 93, 107
Stachel, John, 8, 15, 22
stellar collapse, 155
Synge, John, on general relativity, 155–156,
     177

tensor calculus, 10, 57, 63, 67, 75, 85
tensor multiplication, 63, 65, 73, 193–195;
     and "contraction" of a mixed tensor, 1945;
     examples of, 194–195; external (outer)
     multiplication, 59, 65, 193, 194; inner
     multiplication, 63, 194; mixed multiplica-
     tion, 63, 65, 194
tensors, 57, 61, 75, 191; antisymmetric
     tensors, 61, 193, 205, 206 193; covariant
     tensors, 59, 67, 81, 192, 193, 194, 195,
     204, 208; divergence of a mixed tensor,
     87, 206–207; formation of by differenti-
     ation, 79, 200–203; mixed tensors, 192,
     194; production of new tensors from
     given tensors through differentiation, 79;
     symmetric tensors, 61, 192, 213; tensor
     contraction, 63, 164; tensors of second
     and higher ranks, 191–193. See also
     Einstein tensor; energy-momentum ten-
     sor; metric (fundamental), tensor; Ricci
     tensor; Riemann tensor; tensor calculus;
     tensor multiplication

unified field theory of gravitation, 153
unimodular transformations, 71, 91
universe: expanding, 153–155, 163; static, 151
University of Zurich, 8

variational principle (formalism), 26, 28,
     97, 105, 131, 133, 160, 161, 162; and the
     field equations of gravity and matter,
     229–230. See also Hamiltonian function;
     Lagrangian function
vectors: 57, 75; contravariant vectors, 57,
     59, 190–192; covariant vectors, 57, 79,
     81, 191–192, 203; covariant and con-
     travariant vectors in special relativity,
     113; "curl" of a covariant vector, 206;
     divergence of, 85, 87, 206; "extension"
     of a covariant vector, 204; parallel
     displacement/transport of, 77, 81, 89;
     rotation (curl) of, 85. See also six-vector
volume element, the, 69, 71
von Laue, Max, 115, 160, 177

Weyl, Hermann, 23, 81, 133, 177
Wien, Wilhelm, 2

Zangger, Heinrich, 12, 107
Zurich Notebook (Einstein), the, 10–11, 24,
     28, 29, 69, 73, 83, 91, 107